03-SW-3A

DECENTRALIZED STORMWATER CONTROLS FOR URBAN RETROFIT AND COMBINED SEWER OVERFLOW REDUCTION: PHASE II

by:

Neil Weinstein, P.E., R.L.A., AICP
The Low Impact Development Center, Inc.

Jennifer Cotting
Dan Nees
University of Maryland Environmental Finance Center

Sara Downing
Jeff Lee, ASLA
Benjamin Tauber
Lee+Papa and Associates

Ann English, R.L.A., ASLA, LEED AP
Christopher Kloss
The Low Impact Development Center, Inc.

Charles Glass, Ph.D.
ETEC, L.L.C.

Wayne C. Huber, Ph.D., P.E.
Tracy Morgan
Oregon State University

2009

WERF

The Water Environment Research Foundation, a not-for-profit organization, funds and manages water quality research for its subscribers through a diverse public-private partnership between municipal utilities, corporations, academia, industry, and the federal government. WERF subscribers include municipal and regional water and wastewater utilities, industrial corporations, environmental engineering firms, and others that share a commitment to cost-effective water quality solutions. WERF is dedicated to advancing science and technology addressing water quality issues as they impact water resources, the atmosphere, the lands, and quality of life.

For more information, contact:
Water Environment Research Foundation
635 Slaters Lane, Suite 300
Alexandria, VA 22314-1177
Tel: (703) 684-2470
Fax: (703) 299-0742
www.werf.org
werf@werf.org

This report was co-published by the following organization. For non-subscriber sales information, contact:

IWA Publishing
Alliance House, 12 Caxton Street
London SW1H 0QS, United Kingdom
Tel: +44 (0) 20 7654 5500
Fax: +44 (0) 20 7654 5555
www.iwapublishing.com
publications@iwap.co.uk

Library of Congress Catalog Card Number: 2007943280

IWAP ISBN: 978-1-84339-353-5/1-84339-353-0

This report was prepared by the organization(s) named below as an account of work sponsored by the Water Environment Research Foundation (WERF). Neither WERF, members of WERF, the organization(s) named below, nor any person acting on their behalf: (a) makes any warranty, express or implied, with respect to the use of any information, apparatus, method, or process disclosed in this report or that such use may not infringe on privately owned rights; or (b) assumes any liabilities with respect to the use of, or for damages resulting from the use of, any information, apparatus, method, or process disclosed in this report.

The Low Impact Development Center, Inc.; ETEC, L.L.C.; Lee+Papa and Associates; Oregon State University; and the University of Maryland Environmental Finance Center

The research on which this report is based was developed, in part, by the United States Environmental Protection Agency (EPA) through Cooperative Agreement No. CR-83155901-1 with the Water Environment Research Foundation (WERF). However, the views expressed in this document are solely those of The Low Impact Development Center, Inc.; and neither EPA nor WERF endorses any products or commercial services mentioned in this publication. This report is a publication of WERF, not EPA. Funds awarded under the Cooperative Agreement cited above were not used for editorial services, reproduction, printing, or distribution.

This document was reviewed by a panel of independent experts selected by WERF. Mention of trade names or commercial products does not constitute WERF nor EPA endorsement or recommendations for use. Similarly, omission of products or trade names indicates nothing concerning WERF's or EPA's positions regarding product effectiveness or applicability.

ACKNOWLEDGMENTS

Report Preparation

Principal Investigator:
Neil Weinstein, P.E., R.L.A., AICP
The Low Impact Development Center, Inc.

Project Team:
Jennifer Cotting
Dan Nees
University of Maryland Environmental Finance Center

Sara Downing
Jeff Lee, ASLA
Benjamin Tauber
Lee+Papa and Associates

Ann English, R.L.A., ASLA, LEED AP
Christopher Kloss
The Low Impact Development Center, Inc.

Charles Glass, Ph.D.
ETEC, L.L.C.

Wayne C. Huber, Ph.D., P.E.
Tracy Morgan
Oregon State University

Project Subcommittee

Larry S. Coffman, Chair
President, LNSB, LLLP Stormwater Services Group

Richard R. Horner, Ph.D.
University of Washington

Rick Howard, P.E.
City of Orlando

Amy S. Leib
Philadelphia Water Department, Office of Watersheds

Michael Sweeney, Ph.D., P.E.
EMA, Inc.

Malcolm Walker, P.E.
Larry Walker Associates

Water Environment Research Foundation Staff

Director of Research: Daniel M. Woltering, Ph.D.
Senior Program Director: Jeff Moeller, P.E.

ABSTRACT AND BENEFITS

Abstract:

Phase I of this project demonstrated the technical feasibility of using decentralized stormwater controls in urban areas for retrofits and controlling combined sewer overflows. This technical feasibility was illustrated by a number of early adopters using decentralized controls to complement their existing municipal stormwater and wastewater infrastructure. However, institutional and programmatic issues required further study to broaden the use of a distributed, decentralized stormwater approach. This research evaluates implementation strategies for incorporating decentralized controls into an infrastructure management system. The distributed nature and multiple environmental benefits of decentralized controls necessitate an integrated and inter-departmental management approach.

The researchers identify various implementation strategies for incorporating decentralized controls into urban infrastructure management programs. Using case studies and programmatic and regulatory examples, the report details insight into how best to expedite the adoption of decentralized controls. In the context of distributed controls, the researchers explore managing infrastructure by limiting demand. In addition, they evaluate of economic methods appropriate for assessing environmental costs and benefits, to more fully capture the financial consideration of decentralized controls. The report looks at four commonly used stormwater models and offers guidance for modeling decentralized controls.

Benefits:

- Provides a rationale for a distributed approach to infrastructure management and identifies how decentralized controls address multiple environmental objectives.
- Presents multiple implementation alternatives under which decentralized controls can be integrated or retrofit into the urban environment.
- Provides templates of decentralized control design options according to land use to illustrate the versatility and adaptability of the approach.
- Evaluates multiple economic evaluation methods to more fully and accurately assess the costs and benefits of vegetated, distributed infrastructure.
- Presents guidance on how to appropriately incorporate decentralized controls into stormwater models to accurately reflect anticipated performance.

Keywords: Infrastructure management, green cities, integrated stormwater control.

TABLE OF CONTENTS

LIST OF TABLES

LIST OF FIGURES

LIST OF ACRONYMS

BASMAA	Bay Area Stormwater Management Agencies Association
BMP	best management practices
CSO	combined sewer overflow
GAR	green area ratio
HSPF	EPA Hydrological Simulation Program
LCC	lifecycle cost
LCCA	lifecycle cost analysis
LEED	Leadership in Energy and Environmental Design
LID	low impact development
LTCP	long-term control plan
MS4	municipal separate storm sewer system
NAAQS	National Ambient Air Quality Standards
NPDES	National Pollutant Discharge Elimination System
NRDC	Natural Resources Defense Council
SWMM	U.S. EPA Storm Water Management Model
TEV	total economic valuation
TMDL	total maximum daily load
TR-55	Natural Resource Conservation Service Technical Release 55
USDA	United States Department of Agriculture
U.S. EPA	United States Environmental Protection Agency
UFORE	urban forest effects
USGS	United States Geological Survey
WASA	District of Columbia Water and Sewer Authority
WERF	Water Environment Research Foundation
WWHM	Western Washington Hydrology Model

EXECUTIVE SUMMARY

This report represents the second phase of a research effort aimed at assessing the use and role of decentralized stormwater controls for urban retrofit and combined sewer overflow (CSO) management. The Water Environment Research Foundation Phase I report (03-SW-3, 2006) focused on the technical performance and feasibility of decentralized controls in urban applications.

Much of the initial research focused on the building blocks required to assess the appropriateness of this emerging approach for urban construction and retrofit applications. Specifically, the Phase I report examined the hydrologic performance (e.g., retention, infiltration, evapotranspiration) of appropriate urban decentralized controls and their respective effectiveness at reducing runoff volumes, peak flows, and pollutant concentrations.

Phase II researchers had two objectives:

- Provide additional research into institutional and programmatic issues to complement the Phase I report.
- Present solutions to the challenge of broadening adoption of decentralized controls and enhanced operational methods.

The Phase II findings complete the research team's assessment of urban decentralized controls.

ES.1 Whole Systems Approach

The traditional solution to stormwater management has been a centralized end-of-pipe approach. Understanding how best to use decentralized controls involves considering the role that they will play in the urban environment. Compared to conventional grey infrastructure, decentralized controls provide a more sustainable urban design. Because decentralized controls rely heavily on vegetation or practices that simulate elements of the natural hydrologic cycle, the environmental benefits gained extend beyond stormwater and water quality. There is also a growing "green city" movement as municipalities recognize and seek to capitalize on the many benefits of incorporating green space and green design into the urban environment. Decentralized stormwater controls can play a productive role in this movement and in addressing a number of environmental issues confronting urban areas, including the following:

- urban heat island effect;
- air quality;
- energy;
- climate; and
- aesthetic and community benefits.

Decentralized controls offer a truly integrated approach at comprehensive environmental management. Their unique qualities alter the traditional approach to pollution abatement and present a multi-media/multi-benefit solution. But this new strategy also presents a number of challenging institutional barriers.

Decentralized Approach: The inherent flexibility of decentralized controls makes them a critical component of new urban design focused on functional and sustainable design and landscapes. The performance capabilities of decentralized best management practices (BMPs) are well suited for the many environmental and regulatory requirements that face municipalities including total maximum daily loads, CSO long-term control plans and consent decrees, and other court orders and legal actions. Decentralized controls can be, and have been, applied in a number of ways to optimize infrastructure performance and meet regulatory and natural resource goals. They can be incorporated into a runoff management program in several ways, including:

- stand alone systems;
- hybrid systems; and
- complements to existing grey infrastructure.

Conventional urban BMPs are most often installed as inlet or end-of-pipe units that provide settling or filtration. Decentralized controls provide a range of options before stormwater enters the collection system. Table ES-1 provides an overview of BMP choices with regard to land use.

Table ES-1. BMP Feasibility According to Land Use.

BMP	Commercial, Industrial & Public Buildings	Multi-Family Residential	Single Family Residential	Transportation Corridors	Parks & Open Space
Downspout disconnection	✓	✓	✓		
Filter strips	✓	✓		✓	✓
Infiltration practices	✓	✓		✓	✓
Pocket wetlands			✓	✓	✓
Porous pavement	✓	✓	✓	✓	✓
Rain barrels/cisterns	✓	✓	✓		
Rain gardens	✓	✓	✓	✓	✓
Soil amendments	✓	✓	✓	✓	✓
Tree box filters	✓			✓	
Vegetated roofs	✓	✓			
Vegetated swales	✓	✓	✓		✓

ES.2 Appropriate Locations for Use

A decentralized control approach can be used in all green approaches to community development to improve their stormwater control results. Specific BMPs or control approaches may or may not be applicable depending on site conditions, but the strategy of decentralized stormwater controls is feasible in nearly every situation. Retrofits with decentralized controls are possible in

- areas of highly connected impervious surfaces;
- aged infrastructure; and
- limited impervious or open areas.

Integrated Planning and Design: The coordinated effort of technical and planning staffs and elected officials is critical to the success of a distributed, yet integrated system. Decentralized systems are adaptable to both micro- and macro-scale applications to solving water quality problems. They are able to meet stormwater management goals as well as other planning and design needs for urban settings to the benefit of urban dwellers and downstream recipients of stormwater.

Integrating decentralized controls into an urban environment is accomplished by using structural and non-structural tools. Examples of non-structural tools are included in Table ES-2 and a planning process approach is shown in Figure ES-1.

Table ES-2. Non-Structural Tools for Water Quality Protection.

Non-Structural Tools	Function
Comprehensive master planning	▪ Provides a coordinated approach with consideration of water quality protection.
Stream and wetland setbacks	▪ Place limits on development activities within certain distances of streams and wetlands. ▪ Maintain natural flood and erosion control functions. ▪ Protect infrastructure and homes from moving streams. ▪ Use protection zones. • Based on public health and safety services. • Establishes minimum setback widths. • Details permitted and prohibited structures and uses. • Allows for non-conforming structures and uses. • Includes variance section to maintain buildability and provide flexibility.
Conservation design	▪ Allows for partial conservation of natural resources while development occurs. ▪ Allows homes to be closer together while remainder of development area is in open space. ▪ Requires understanding of site's soils, topography and natural features.
Distributed stormwater management	▪ Stormwater controls are site features implemented at the source of impacts. ▪ Potentially reduces stormwater infrastructure costs. ▪ Promotes public participation and public education in pollution prevention and maintenance stormwater management practices. ▪ Satisfies NPDES Phase II stormwater requirements. ▪ Application for retrofit projects.

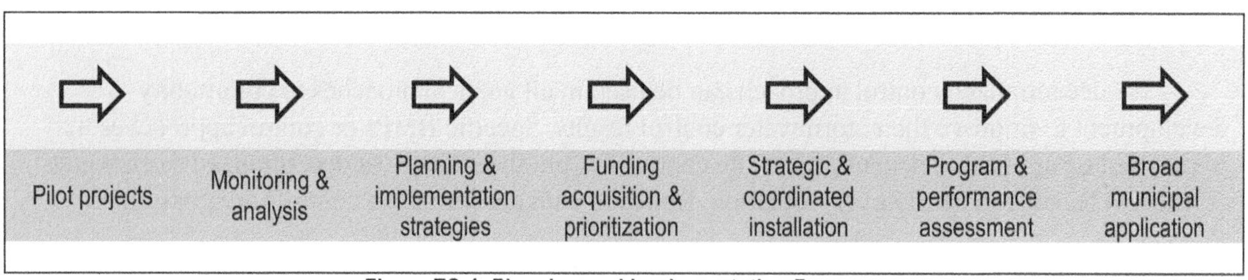

Figure ES-1. Planning and Implementation Process.

U.S. EPA advocates a watershed-based planning model because the overlapping political issues are best reconcilable at the watershed level. This approach requires technical and institutional resources to fully develop an integrated approach that uses decentralized controls. Environmental goals will need to be identified to assess measurable and meaningful improvements to water quality.

Implementation Alternatives: Each separate stormwater or combined sewer system has unique operational and system characteristics that preclude prescriptive guidance. A few planning and implementation scenarios are the most likely to be used for decentralized control installation and selection. Four likely approaches to encouraging decentralized control inclusion are

1. pilot installations by an agency or organization at any level;
2. watershed-based approaches, which will typically be overseen by a regional- or higher-level authority;
3. building and site redevelopment projects, which work to implement distributed controls as part of their building process; and
4. large-scale urban revitalization projects, which will often be a collaborative process between agencies at a variety of levels, including private developers.

For each municipality, the particular stormwater drivers will need to be identified. This will help determine which controls are appropriate for the land use and critical stormwater parameters (e.g., volume reduction primarily in combined sewer systems, volume reduction and water quality in areas with sensitive receiving waters). BMP identification should be undertaken parallel to determining funding sources for implementation and monitoring protocols for oversight of the final control system.

Fiscal Benefits of Decentralized Controls: Decentralized control measures have the potential to influence the financial costs of mitigating CSOs and stormwater runoff. The costs associated with stormwater management programs are directly related to the affect of development practices on nature's ability to maintain a water balance. Distributed BMPs can help reduce the costs associated with stormwater management by restoring this balance, thereby decreasing the amount of pollution introduced into waterways and relieving the strain on stormwater and wastewater infrastructure.

Implementing decentralized control programs may be the most cost effective and efficient way to achieve water quality goals and requirements. However, the multiple benefits provided by decentralized controls and the ability to capture the asset values of ecosystem services necessitates more comprehensive methods of economic valuation than those traditionally used to assess stormwater and CSO programs. Table ES-3 presents various methods for evaluating the economic considerations when implementing decentralized controls.

Table ES-3. Comparison of Costing Methods.

Method	Description	Positives	Drawbacks
Damage cost avoided, replacement cost, and substitute cost methods	• Estimates values of ecosystem services based on the cost of avoiding damages due to lost services, replacing ecosystem services, or providing substitute services.	• Most useful and appropriate when actual damage or avoidance costs can be measured.	• Do not provide strict measures of economic values, but rather assume that avoided or replacement costs provide useful estimates of value.
Life-cycle cost analysis	• Considers all costs from planning and design through disposal.	• Good method for evaluating competing alternatives that have different initial costs, O&M costs, and life spans.	• Subjective nature of determining life span complicates analysis.
Benefit-cost analysis	• Considers benefits gained from investment as well as costs.	• More accurately reflects the full range of environmental and economic benefits and costs.	• Valuation of community and environmental benefits currently inadequate.
Productivity method	• Estimates the economic value of ecosystem services or products that contribute to commercially marketed goods.	• Can effectively estimate costs when treatment costs of environmental goods (e.g., water) are known.	• Only estimates the value of commercially marketable products.
Hedonic pricing method	• Estimates economic values for environmental services that directly affect market prices.	• Useful for estimating property value changes from use of decentralized controls.	• Limited to assessing affect on market prices.

ES.3 Stormwater Modeling Approach

Whatever processes are used to integrate decentralized controls into capital projects, development projects, and/or planning, a stormwater model capable of simulating decentralized controls will be an essential part of any stormwater management program. Stormwater modeling is needed for defining the runoff problem, generating and evaluating alternatives, decision making, and long-term management.

Modeling decentralized controls uses the same principles and processes as modeling conventional systems. An obvious realization is that instead of one BMP, several will be used, hinting that the scale of a decentralized modeling project can quickly expand. Ideally, the model must be capable of simulating large decentralized systems. Providing complex flow routing, explicitly addressing specific unit processes during each water balance calculation, and being able to physically represent the BMP structure with a high level of accuracy are also characteristics that an ideal model would exhibit for decentralized modeling. For this report, researchers examined four models: U.S. EPA's Storm Water Management Model, version 5 (SWMM5), NRCS's Technical Release 55 (TR-55), U.S. EPA's Hydrologic Simulation Program - FORTRAN model (HSPF), and Western Washington's Hydrology Model, version 3(WWHM3).

SWMM's long history and capability has secured it as a popular choice in BMP design. SWMM's inherent ability to route flow from an impervious surface to a pervious surface (and from one subcatchment to another) makes it an ideal candidate for simulating decentralized controls. It is also the most comprehensive regarding the simulation of infiltration, as it applies several well-known techniques and allows differentiation between BMPs.

TR-55 is perhaps the least capable of the four in this report for modeling decentralized controls. The most crucial unit processes, infiltration and evaporation, are consolidated into the initial abstraction term, and ultimately put a large emphasis on the curve number. Watershed planners may find it viable to get a broad, general idea of a watershed's developmental response and/or problem areas based solely on water quantities.

The HSPF model is geared more toward water quality analyses rather than water quantity. HSPF is most commonly used not as a BMP design package, but instead as a tool to assess the water quality effects of BMPs as well as considerable land-use change on a watershed scale. Its strengths lie in its ability to account for the fate of multiple contaminants during point and non-point pollution sources through continuous simulation. It is able to produce peak flows, volumes and hydrographs, and account for the vertical water balance in ways that are very useful for simulation of distributed controls.

WWHM3, although new, seems to be a very promising addition to the crowded field of hydrologic models. The model is proving to be a useful and diverse tool for many private consulting firms across western Washington. The model's flexibility has allowed it to be used for hydraulic purposes outside of its original design, as it has been applied to culvert sizing and wetland hydro-period analyses.

Designating which model is more suitable for simulating decentralized controls is difficult without actual data testing. A likely test would include a comparison of each model's behavior in an aggregated versus disaggregated simulation, singling out the model with the least margin of error. It would also be beneficial to see, using the same data set, the computed reduction in runoff each model produces when implementing the same decentralized controls. Comparing the results of these two tests would provide much more insight into the limitations of each model.

CHAPTER 1.0

INTRODUCTION

1.1 Background

Phase I researchers focused on the technical performance and feasibility of decentralized controls such as vegetated roofs and swales and soil amendments in urban applications (Water Environment Research Foundation report 03-SW-3, 2006). Specifically, the researchers examined the hydrologic performance (e.g., retention, infiltration, evapotranspiration) of appropriate urban decentralized controls and their respective effectiveness at reducing runoff volumes, peak flows, and pollutant concentrations. In addition, the report evaluates the technical issues and the feasibility of incorporating these controls into an urban environment.

The Phase I researchers found that several cities adopted these measures and were using them with success. Individual municipalities use varied approaches and strategies and a mix of different technologies, but the majority of applications were successful at reducing runoff volumes (and consequently peak flow rates) and pollutants.

Although establishing technical feasibility was a critical first step in validating decentralized controls as a viable stormwater control measure, researchers still needed an assessment of institutional and programmatic issues. The challenge of broadening adoption of decentralized controls involved revising technical practices and operational methods.

In this Phase II report, the researchers address these broader themes to assess the use and role of decentralized stormwater controls for urban retrofit and combined sewer overflow (CSO) management. Taken together, Phase I and II reports fully explore issues involved with managing stormwater in a distributed fashion.

1.2 A Whole Systems Approach

Stormwater management has long resided within the engineering domain; largely as a conveyance and waste management issue. The prevailing concerns included flooding prevention, and protection of property and life. The dominant approach involved conventional infrastructure practices and hydraulic considerations. Although decentralized controls are a viable technical option, the largest barriers involve institutional inertia and changing the "business model" for stormwater.

Managers must consider the role that these decentralized controls will play in the urban environment to best use these controls. Compared to conventional grey infrastructure—which is largely single-function—decentralized controls provide a layered approach to environmental management for stormwater control and greener, more sustainable urban design. Because decentralized controls rely heavily on vegetation or practices that simulate elements of the natural hydrologic cycle, the environmental benefits gained extend beyond stormwater and water quality. There is also a growing "green city" movement as municipalities recognize and seek to capitalize on the many benefits of incorporating green space and green design into the urban environment. Decentralized stormwater controls can play a productive role in this movement and in addressing a number of environmental issues confronting urban areas.

- ***Urban heat island*** In cities, hard, reflective surfaces absorb solar radiation and re-radiate it as heat to create an urban heat island. Summer temperatures in urban areas can average 10° F higher than suburban temperatures. Decentralized controls counteract urban heat island impacts by substituting soils and vegetation for hard, heat-absorbing materials common in urban areas, or by using lighter colored, more reflective alternatives (e.g., permeable paving blocks). Vegetation creates shade and emits water vapor—both of which cool hot air. Water vapor emitted by plant material cools ambient temperatures because heat energy is used by vegetation to evaporate water.

  A surface's radiation absorbing potential is measured by its albedo, the ratio of radiation reflected from a surface to the incoming radiation onto the surface. The lower a surface's albedo, the more radiation it absorbs. Vegetation reflects more radiation because grasses, deciduous plants and trees, and coniferous trees have albedos that are, on average, two to three times greater than those of asphalt, tar, and gravel that are typical urban construction and building materials. (Grant et al., 2003).

- ***Air quality*** Higher temperatures common in urban areas can translate to lower air quality conditions because increased temperatures can increase ground-level ozone concentrations. Hydrocarbons and nitrogen oxides react to form ground-level ozone in a process catalyzed by the ultraviolet radiation in sunlight. Higher temperatures tend to correlate with higher ozone (O_3) concentrations in part because higher temperatures increase reaction rates (Vingarzan and Taylor, 2003). Urban air quality is also affected by industrial emissions and mobile source pollution.

  Decentralized controls improve air quality through several mechanisms. Vegetation reduces urban temperatures, which decreases the reaction rates involved in the creation of ground-level ozone. Vegetation also improves air quality by physically filtering air pollutants. Leaf stomata absorb gaseous pollutants and plant surfaces can adsorb particulate matter. Adsorbed particles can remain adhered to the plant, be absorbed into the plant material, washed off to the soil, or redispersed to the atmosphere. The U.S. Department of Agriculture (USDA) Forest Service's Urban Forest Effects, or UFORE, model has estimated that the vegetation associated with decentralized controls can remove significant amounts of five common urban air pollutants on an annual basis: O_3; particulate matter less than 10 microns in diameter (PM_{10}); nitrogen dioxide (NO_2); carbon monoxide (CO); and sulfur dioxide (SO_2) (Currie and Bass, 2005).

- ***Energy*** By shading and insulating buildings, green space lowers ambient temperatures, thus decreasing the energy needed for heating and cooling. For example, green roofs absorb and hold thermal energy, and release the energy when the surrounding ambient temperatures cool. Acting as heat sinks, green roofs prevent the wide swings in surface temperatures experienced by conventional roofs and reduce the heating and cooling demand on buildings (Miller, 2004). Estimates suggest that green roofs can reduce the heating and cooling costs of a building 10–15% (American Society of Landscape Architects, 2006). Also, when assessed in aggregate, a system of decentralized controls implemented city-wide could lower ambient summer temperatures because of increased vegetation and reduce overall cooling demands.

- ***Climate*** Decentralized controls such as green roofs and tree box filters provide a method of addressing climate change. Decreasing the energy needed for heating and cooling can reduce carbon dioxide (CO_2) emissions when power is provided by fossil fuel-fired power plants. The vegetation used with decentralized controls also provides carbon storage. The respiratory process of plants converts atmospheric CO_2 into cellular carbon. Mature vegetation stores ever greater amounts of carbon.

  Decentralized controls can help to mitigate the impacts of increased temperatures by counteracting the urban heat island. The cooling impacts of decentralized controls can also help to offset declining air quality conditions that will likely accompany higher temperatures. In addition, as temperatures increase, evapotranspiration rates are likely to also increase, potentially increasing the water retention capabilities of decentralized controls. Vegetated systems are dynamic systems that respond to external environmental stimuli. Additional vegetative growth and increased rates of evapotranspiration expected with warmer temperatures may counteract some of the forecasted warming and precipitation impacts.

- ***Aesthetic and community benefits*** Trees and plants improve urban community livability by providing recreational and wildlife areas. Studies in Philadelphia have demonstrated that new tree plantings increased surrounding housing values up to 10% and that vacant lot improvements showed up to a 30% increase in surrounding housing values (Wise, 2007). This has both private and public benefit: the private value gained by homeowners is translated into additional public tax revenue from higher assessments.

1.3 Integrating Programs

Historically, consideration and treatment of pollution sources has been segregated by media (e.g., air or water). A regulatory structure that has addressed air and water singularly, for instance, has largely resulted in singularly focused compliance efforts. This is the case with the overflow control policy, but adopting decentralized controls as a means of control can provide multi-media compliance. Improving air quality is an example.

The Clean Air Act establishes National Ambient Air Quality Standards (NAAQS) for six criteria pollutants: ground-level ozone (O_3), nitrogen oxides (NO_x), carbon monoxide (CO), fine particles, sulfur dioxide (SO_2), and lead. Each state is responsible for establishing emission standards that will result in attainment and maintenance of the NAAQS and submitting a state implementation plan that demonstrates how the NAAQS will be achieved. To meet standards states have primarily regulated the emissions of existing sources of criteria pollutants and federal New Source Performance Standards regulate new sources or modifications to existing sources

(Salzman and Thompson, 2003). Many of these efforts have focused on technological pollution control improvements or operational modifications to limit air pollution.

Even with these efforts, many urban areas fail to meet air quality standards. The two criteria pollutants for which urban areas are most often in non-attainment are particles and ground-level ozone. Continued non-attainment can have significant impacts beyond health impacts to the public. The federal government may withhold federal highway funding for states and municipalities that do not demonstrate significant progress toward achieving attainment.

Air pollution is a complicated regional problem and many municipalities do not have control over the sources of pollution that impact their area. Also, urban factors, such as higher temperatures, contribute to air quality problems. The occurrence of code orange or red air quality days generally corresponds with the hottest days of the summer.

Municipalities take several approaches to address poor air quality days. Free rides on public transportation and discouraging the public from refilling gas tanks or using gasoline powered lawnmowers are programs used to try to improve air quality. Municipalities have also purchased natural gas or hybrid buses to help limit emissions from the municipal fleet. Decentralized controls also offer an additional measure that can improve air quality through the various methods described previously. When integrated in a holistic and coordinated manner, decentralized controls offer the potential of improving atmospheric conditions within the urban environment. To achieve significant benefit, decentralized controls would need to be incorporated across large sections of a municipality but a potential exists to integrate water and air pollution control efforts.

By altering the traditional approach to pollution abatement and presenting a multi-media/multi-benefit solution, decentralized controls offer an integrated approach at comprehensive environmental management. But this approach also presents a number of challenging institutional barriers. Decentralized controls can be implemented in the same traditional manner as other water infrastructure projects with a public works department employing them as a new set of stormwater best management practices (BMPs). Interagency or departmental approaches will most efficiently optimize decentralized controls, both environmentally and economically. A public works department may have a narrowly focused regulatory or administrative mission of managing stormwater, but the non-water quality environmental and community benefits of decentralized controls indicate a need to move outside of traditional stovepipes and operate in a coordinated fashion across a governmental structure. This is an institutional challenge from an operational and funding perspective as budgets and management responsibilities have traditionally been segregated by the bright lines that exist between departments. However, the extended operational and performance capabilities of decentralized controls merit a re-evaluation of inter-department coordination.

Decentralized controls offer the path to a new urban model of environmental performance aimed ultimately at enhancing the sustainability of urban areas. This approach has the potential to transform the view of urban areas as removed and separate from the natural environment and as having lower environmental quality. The "greener" approach provided by decentralized controls will aid municipalities in meeting their regulatory requirements but also enhance the natural conditions of the community. The ability to simultaneously address multiple environmental concerns offers a new method of environmental protection.

The decentralized approach promotes economic as well as environmental sustainability. Compliance with environmental regulatory requirements presents a financial challenge to municipalities balancing competing requirements. Investment in decentralized controls optimizes the financial expenditures because they provide benefit across media and offer many beneficial outcomes. These approaches optimize and enhance the performance of the existing grey infrastructure. Municipalities have heavily invested in traditional infrastructure; therefore its continued service is a critical concern. The current estimates for wastewater infrastructure needs exceed $200 billion and funding the repairs and improvements needed to extend the operation of the thousands of wastewater systems is a fiscal challenge (U.S. EPA, 2008). Decentralized controls complement grey infrastructure by reducing the burden and demand on the existing systems and optimizing their performance. The incremental funding of a distributed system for stormwater management can also limit the required debt service that is needed for large-scale capital improvements.

The Phase II researchers extended the technical considerations of the Phase I effort and evaluated the institutional and financial implications of adopting the distributed approach of decentralized controls. In addition, their report focuses on how best to apply decentralized controls to urban land uses. The advance of decentralized controls primarily hinges on the institutional response to this new management approach and the performance of the controls Modifying the centralized treatment paradigm that has predominated stormwater management requires an institutional coordination that has not often been needed previously.

CHAPTER 2.0

CONCEPTS AND POLICIES

2.1 Building the Case for Decentralized Systems

Managing stormwater for quality is a fairly recent development. Although managed at some level from the beginning of urbanization, historical stormwater control focused on expediting the flow to local rivers and waterways. Minimizing loss of life, property damage, and nuisance flooding were primary concerns to early stormwater managers. Open drains dominated the early urban landscape; with widespread adoption of indoor plumbing, the existing open drains transported wastewater with the stormwater. Public health considerations drove the drainage systems into pipes underground. The capacity of the combined sewers became a critical concern because wastewater treatment plants were installed as an initial step to protect water quality. Overflows from combined sewers were installed to relieve hydraulic limitations from within the collection system or at the downstream treatment plant.

The evolution of combined sewer systems is informative when assessing the most efficient means of reducing overflows. This evolution was not necessarily deliberate but rather reactive to external pressures that emerged in response to increasing public health and environmental awareness. However, this responsive mindset has persisted as the prevailing conventional wisdom for reducing overflows. Responses to the issue have focused almost exclusively on adding additional capacity in a reflexive manner. The dominant paradigm has been: if there is a capacity problem, add capacity. This approach will improve system performance but it will likely not satisfactorily remedy the problem because it treats the symptoms, but not the cause.

This one-dimensional approach is limited in its potential effectiveness because it does little to address the upstream stormwater issues that are responsible for the vast majority of overflows. Foremost amongst these issues is the increased development that adds ever more stormwater to the collection system. The adequacy of water infrastructure, and infrastructure in general, has most often been judged by its capacity. Capacity has been viewed through the single prism of supply; that is, whether the infrastructure is capable of delivering the necessary volume. In terms of stormwater and wastewater infrastructure, does the pipe system supply the needed volume and flow rate to prevent overflows. Little attention has been given to demand management, or limiting the necessary amount of capacity. Decentralized stormwater controls are a demand management approach. Using decentralized controls eliminates a percentage of the stormwater volume entering the collection system and "creating" capacity by lowering system requirements.

A corollary can be found in the electric utility industry, as stormwater and combined sewer systems are often managed as utilities in urban areas. During the 1970s, Seattle city planners anticipated electricity demand would continue to increase according to historical norms and double every 10 years, resulting in a capacity shortage. Seattle's mayor and superintendent of Seattle City Light wanted to meet this demand by purchasing a 5% share in two new nuclear plants being built in the region, providing increased capacity to meet the projected shortage. In 1976, the city council voted against obtaining power from the new nuclear plants and passed five resolutions adopting conservation as the long-term energy strategy (Wilma, 2001). The decision was based on an economic rather than an environmental rationale. City residents and the council felt that conserving energy would be equivalent to adding additional capacity but could be achieved at a lower cost. The city invested in distributed, small-scale conservation measures that. produced results similar to adding capacity, and prevented new power plant construction for 20 years, at 20% of the cost for new generation (Wise, 2007). These initiatives built on the earlier success of Seattle's "Kill-a-Watt" program, which was adopted in July 1973 as a means of reducing electricity demand until additional capacity could be supplied. The success of the program was immediate, with voluntary measures reducing consumption 7% by the end of the year (Wilma, 2000).

Today, demand management in the electric utility industry is common. When confronted with the tremendous capital cost and lag time for building new capacity, electric utilities often promote and sponsor end-use energy conservation measures. The realization that conservation can substitute for additional capacity and with additional economic and environmental benefits has been internalized within the utility and public decision-making process. This conceptual approach is applicable to stormwater and combined sewer management but has yet to be widely institutionalized. Some jurisdictions have adopted this approach and, it is no surprise that Seattle is at the forefront of decentralized stormwater control efforts.

Demand management in stormwater differs a bit in application from that in the electric utility industry. Enhancing capacity within an electric utility by assessing demand primarily looks at end-uses and the conservation measures that can be instituted to limit the amount of electricity "requested" from the utility. Limiting demand on stormwater or wastewater infrastructure is the opposite side of the spectrum; reducing upstream loads and the conveyance demand on the collection and treatment system. This approach requires a significant shift in the normal approach applied to water infrastructure and altering how sufficient capacity is assessed.

Upgrades to stormwater and combined sewer systems, like other utilities, are capital-intensive projects. U.S. EPA's latest Clean Watersheds Needs survey estimates that $9 billion is needed for stormwater management and nearly $55 billion is needed for CSO correction (in 2004 dollars) (U.S. EPA, 2008). This presents a significant economic burden, whereas the lack of adequate infrastructure capacity presents a significant public health and environmental burden. A demand management approach with decentralized controls provides an opportunity to alleviate aspects of the economic and environmental impacts.

2.2 A Decentralized Approach

Applying decentralized controls requires a different model of assessing urban management opportunities. Conventional controls are largely buried, unseen, and given little thought by anyone outside of the utility responsible for their operation and maintenance. Decentralized controls are woven throughout the urban landscape and, therefore, necessitate a different implementation approach, including comprehensively assessing the urban landscape. Retrofitting stormwater controls into a space-limited urban area is complicated by existing utilities, compacted and contaminated soils, right-of-ways, and maintenance access. One of the strengths of decentralized controls is the multitude of available BMPs that may be used and the manner in which treatment trains may be constituted.

Decentralized controls are inherently flexible, which make them a critical component of new urban design focused on functional and sustainable design and landscapes. The performance capabilities of decentralized BMPs are well suited for the many environmental and regulatory requirements that face municipalities including total maximum daily loads (TMDLs), CSO long-term control plans (LTCPs) and consent decrees, and other court orders and legal actions. Applied effectively, decentralized controls optimize infrastructure performance and meet regulatory and natural resource goals.

1. Stand-alone systems – In urban retrofit applications, decentralized controls can be applied as discrete, stand-alone BMPs to meet specific localized water quality needs;
2. Hybrid systems – An application that creates a treatment train of decentralized controls or decentralized and conventional controls. This type of approach is often used for separate sewer systems to provide enhanced water quality before discharge; and
3. Complement to existing pipe (grey) infrastructure – Often employed in combined sewer areas to relieve the hydraulic burden on grey infrastructure and enhance its operational performance.

Utility managers most often install conventional urban BMPs as inlet or end-of-pipe units that provide settling or filtration, with the focus of BMP implementation being suitability with the configuration and hydraulic function of the collection system. In contrast, decentralized controls take place before stormwater enters the collection system, so managers must consider land use, pollutant type and load, building construction, and basin type. Evaluating these factors leads to the selection of optimal decentralized treatment practices and exclusions of others. For example, green roofs are ideal for downtown commercial areas but less so for single-family residential land uses. Table 2-1 provides an overview of BMP choices with regard to land use and Figure 2-1 illustrates decentralized control options for urban residential design.

Table 2-1. BMP Feasibility According to Land Use.

BMP	Commercial, Industrial & Public Buildings	Multi-Family Residential	Single Family Residential	Transportation Corridors	Parks & Open Space
Downspout disconnection	✓	✓	✓		
Filter strips	✓	✓		✓	✓
Infiltration practices	✓	✓		✓	✓
Pocket wetlands			✓	✓	✓
Porous pavement	✓	✓	✓	✓	✓
Rain barrels/cisterns	✓	✓	✓		
Rain gardens	✓	✓	✓	✓	✓
Soil amendments	✓	✓	✓	✓	✓
Tree box filters	✓			✓	
Vegetated roofs	✓	✓			
Vegetated swales	✓	✓	✓		✓

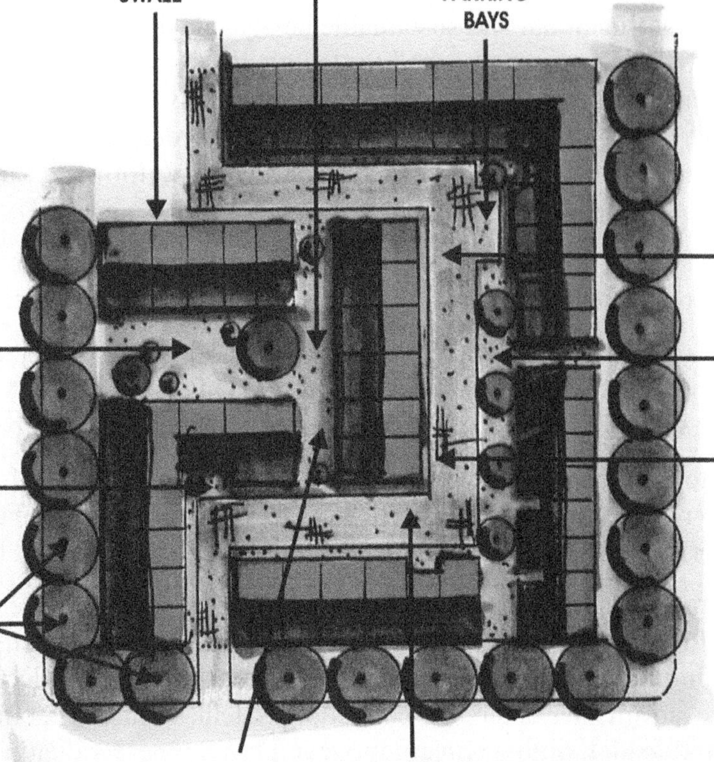

Figure 2-1. Integrated Urban Residential Design with Decentralized Controls.

Widespread implementation of decentralized controls reduces significantly stormwater runoff across a basin or catchment area, but the essential nature of these controls allows for incremental discrete application. In this way, larger water management goals can be accomplished in a step-wise fashion that allows localized issues to be dealt with quickly.

Chicago managers incorporated this stepwise approach with its green alley program, which is used to address site-specific stormwater issues and provide more effective long-term stormwater management. The city has the most public alleyways in the world – with 1,900 miles, constituting approximately 3,500 acres of impervious surfaces – and has targeted these alleys for retrofit with permeable paving to encourage source control and reduce runoff. The program alleviates the immediate effects of neighborhood flooding and addresses a significant source of runoff within the city. Specific standards for design and construction were established to augment the effort with outreach materials including a handbook, video, and training (Wise, 2007). Alleys that were previously a source of stormwater runoff have been converted to treatment systems. Some alleys retain and infiltrate the volume of a 3-inch, 1-hour storm (Chicago Department of Environment, 2005).

2.3 Urban Retrofits

Urban stormwater constitutes a large and growing influence on water quality. Linked to urbanization and development, stormwater runoff has increased as a focus of water quality impairment over the past two decades. Emblematic of the problem is that impervious cover and land use changes are increasing at approximately twice the rate of population growth, meaning that, on a national average, development and its watershed impacts are increasing at a rapid rate (Beach, 2002). Consequently a great amount of attention has been given to managing stormwater from new development.

However, urban areas represent the "past sins" of stormwater management and are a primary cause of stormwater pollution. In urban streams, concentrations of total phosphorus exceeded U.S. EPA's goal for nuisance growth in 70% of streams, insecticides are usually at a higher concentration than in agricultural areas, and fecal coliform bacteria commonly exceed recommended standards for water recreation [U.S .Geological Survey (USGS), 2008]. Considering that a majority of urban areas were developed prior to stormwater controls and that municipal separate storm sewer systems (MS4) are primarily designed for conveyance and not water quality improvement, BMP retrofits present a significant technical and economic challenge. The Phase I and II stormwater regulations require the development of management plans and post-construction controls, but many portions of urban areas will not be affected by these requirements if they are not redeveloping. Emerging TMDL requirements will require that municipalities address these areas to meet permit limits.

As a distributed management approach, decentralized controls offer a viable solution to managing stormwater in the space-limited confines of urban and ultra-urban areas. The volume reduction and pollutant treatment capabilities of decentralized controls provide great opportunity for pollutant load reductions from MS4 discharges. In certain sensitive watersheds, decentralized controls are already playing a significant role in reducing stormwater pollution. In Seattle and Vancouver, British Columbia decentralized controls have been installed in salmon bearing watersheds to reduce stormwater pollutant loads. The Seattle SEA (Street Edge Alternatives) Street has reduced runoff by 99% since the project was installed and has not discharged stormwater since December 2002, reducing the pollutant load from the street to almost zero since project completion (Horner, 2004). Vancouver's Crown Street, completed in 2005 and currently being monitored, was sized for the 10-year storm and is anticipated to retain 90% of the annual rainfall (Steed, undated).

2.4 CSO Federal Policy

Overflows from combined sewer systems are regulated by the Combined Sewer Overflow Control Policy promulgated by U.S. EPA in 1994 (*Federal Register*, 1994). These regulations establish prescribed and certain requirements to demonstrate compliance with federal limitations. This regulatory certainty creates drivers and incentive for municipalities to develop sound overflow control strategies to ensure compliance with permit standards. Decentralized controls, often thought of as a method of residential stormwater control, have been used with greater frequency for urban combined sewer management.

A survey of municipalities with more extensive decentralized control projects finds that many of the more robust programs are being used to reduce stormwater flow into combined sewer systems rather than being used solely as part of the separate stormwater collection system infrastructure as may be expected. Stormwater regulations for MS4s are generally aimed at reducing pollutants. The regulations require use of minimum control measures and do not provide sufficient incentive for the use of an alternative stormwater management approach. The requirements outlined in U.S. EPA's CSO control policy and incorporated into each municipality's LTCP mandate specific performance criteria and targets. The performance requirements and regulatory certainty within the combined sewer program encourages municipalities to look for compliance strategies that are outside the realm of traditional, conventional controls. Simply put, when forced to reduce the number and volume of CSOs, municipalities recognize and appreciate how decentralized controls can reduce or delay stormwater inflow into the combined sewer system.

2.4.1 CSO Control Policy Requirements

The final rulemaking allows a great deal of flexibility for municipalities to select CSO control strategies appropriate for their individual system performance. However, specific requirements that must be met to demonstrate compliance are identified in the rulemaking. Although directly focused on the control of CSOs, U.S. EPA's policy stipulates that the selected LTCP eventually result in compliance with Clean Water Act requirements, including meeting state-defined water quality standards (*Federal Register*, 1994). This requirement that receiving waters eventually achieve designated use standards necessitates, in the long term, that control efforts are comprehensive enough to manage stormwater in a manner that substantially reduces its burden on the combined sewer system. This need will encourage municipalities to adopt decentralized control strategies to enhance the performance of their entire control system.

It is illustrative to look at the requirements for LTCPs developed under the "demonstration approach" to see what is considered full compliance with the CSO Control Policy.[i] Each of these four criteria must be met when developing a LTCP with the demonstration approach (U.S. EPA, 1995).

1. The planned control program is adequate to meet water quality standards and protect designated uses, unless water quality standards or uses cannot be met as a result of natural background conditions or pollution sources other than CSOs;
2. The CSO discharges remaining after implementation of the planned control program will not preclude the attainment of water quality standards or the receiving waters' designated uses or contribute to their impairment. Where water quality standards and designated uses are not met in part because of natural background conditions or pollutions sources other than CSOs, a TMDL, including a wasteload allocation, a load allocation or other means should be used to apportion pollutant loads;
3. The planned control program will provide the maximum pollution reduction benefits reasonably attainable; and

[i] U.S. EPA provides two methods of demonstrating the adequacy of a LTCP: the "demonstrative" or "presumptive" approaches. When attainment of water quality standards will not be met by CSO control alone due to the presence of additional sources of pollution and sufficient data are available to demonstrate that the proposed LTCP will result in an appropriate level of CSO control, the demonstrative approach is used. In instances where data and modeling are unable to predict the level of control necessary to meet water quality standards, the presumptive approach is used to determine an adequate level of control that is protective of water quality standards.

4. The planned control program is designed to allow cost-effective expansion or cost-effective retrofitting if additional controls are subsequently determined to be necessary to meet water quality standards or designated uses.

As opposed to regulations that could have merely dictated technical parameters for the operation of a combined sewer system and the limitation of overflows, these criteria establish a comprehensive set of measures to demonstrate compliance with the federal overflow policy. Because these criteria focus on the quality of receiving waters, selected CSO control measures must be able to not only address the immediate hydraulic deficiencies of the collection system, but also, to the intent of the policy, contribute to a comprehensive watershed management strategy. Although federal regulations are a significant impetus for establishing effective CSO management, they are one piece of a much larger regulatory and planning network that must be coordinated to develop a thorough watershed approach. Table 2-2 provides an overview of the programmatic and institutional influences over CSO programs.

Table 2-2. Planning and Regulatory Influences on CSO Programs.

Governmental Level	Planning, Advisory & Regulatory Agencies	Tools
Federal	• EPA • Department of Transportation • Federal Highway Administration • Nonprofit organizations	• NPDES permits • Grants • Federal funding • Taxes • Tax incentives
State	• Department of Environment / Natural Resources • State Department of Transportation • Department of Housing • Watershed organizations	• Permits • State funds • Bonds • Taxes • Tax incentives
Local	• Public works • Water and Sewer Authority • Planning department • Zoning board	• Permits • Inspections • Fines • User impact fees • Preferential zoning • Local bonds

2.4.2 Performance Criteria Used to Demonstrate Compliance

U.S. EPA's regulatory requirements influence the two predominant methods of conventional CSO control, tunnel and storage systems and sewer separation, in slightly different manners. In practice, LTCPs that select tunnel and storage systems recognize that the selected control measures will not eliminate CSOs but will significantly reduce their occurrence and the volume of discharges. Performance requirements, for instance, may stipulate that wet weather overflows from the combined sewer system be limited to a limited number of events per year and that the volume of overflows be decreased by 95% or more. Criteria such as these create an interesting situation for municipalities.

It is generally economically prohibitive to design storage and conveyance systems that can adequately handle stormwater volumes from storm events of widely divergent intensities and rainfall amounts. Therefore, establishing design criteria that will allow control and treatment of an overwhelming percentage of rainfall events is expedient from a permitting and operational standpoint. However, from a public perception standpoint, any overflow event, especially when overflows may cause beach closures, fish advisories, or other very public consequences, may be seen as an unacceptable health risk. Therefore, municipalities may be faced with a situation in which they essentially must perform beyond compliance with permit standards.

Given the costs of adding additional storage capacity to tunnel systems, decentralized controls provide a cost-effective alternative to further limiting the volume of CSOs by reducing the volume of stormwater entering the collection system. Integrating decentralized controls into urban stormwater control efforts to complement the operation of a tunnel system enhances the performance of the overall system. Decentralized control practices can be integrated into the urban environment over a period of many years, thereby providing an increasing benefit to the infrastructure network. Also important is that decentralized controls, when used as a component of a control program, help to satisfy the fourth criterion of the demonstration approach that the adopted control strategies allow for cost-effective expansion or cost-effective retrofitting.

The approach of integrating decentralized controls into programs with tunnel systems has already been adopted. Chicago, Milwaukee, and Portland (Oregon) combined have spent over $7 billion on deep tunnel systems, but each is installing various decentralized control techniques to enhance the operation of their systems and improve compliance efforts [The Natural Resources Defense Council (NRDC), 2006]. These municipalities recognize that ensuring compliance of their tunnel systems over the long term with future increases in development and population growth and unpredictable rainfall events is a challenge. The use of decentralized controls provides the ability to continually improve the performance of the infrastructure and provide operational flexibility.

Portland's relatively inexpensive downspout disconnection program has been expanded to the point where it currently is removing over one billion gallons of stormwater from the collection system annually. Pilot testing of other decentralized control practices in Portland's residential neighborhoods has indicated that stormwater retention will be significant enough to prevent basement flooding. Similarly, Chicago has also used rain barrels and permeable pavement as a strategy to reduce basement backups in afflicted neighborhoods (NRDC, 2006). In the early part of this decade, Washington, DC, submitted a draft LTCP that proposed building three separate tunnel systems over 20 years. Public comment to the draft recommended incorporating decentralized controls into the plan to provide an alternative means of stormwater management and offer the possibility of potentially reducing the size of two of the tunnels to be built in the out years. The District's final plan identifies that decentralized controls will be included as a control strategy.

The regulatory criteria may have a greater influence on control plans that use sewer separation to address CSOs. Sewer separation is an attractive approach because it can virtually eliminate the flow of stormwater into the wastewater collection system; the trade-off is an increased discharge of untreated stormwater from the separate storm sewer system. Therefore, sewer separation that does not provide for adequate mitigation of stormwater runoff pollution sufficient to achieve water quality standards will fail to meet the first criterion of the overflow control policy. To date, water quality standards and TMDLs have not greatly affected stormwater treatment and control requirements. However, as efforts to improve water quality evolve, these regulatory instruments will not be able to be ignored and will be an increasingly large influence on stormwater management. With this consideration, decentralized controls are an ideal component of management strategies that use sewer separation as a primary control mechanism because of stormwater volume reductions and water quality improvements that can be gained from their use.

2.4.3 Public Involvement and Outreach

U.S. EPA (1995) advocates public outreach in CSO control programs. Traditionally, this has involved municipalities conducting meetings to inform the public of proposed control measures, environmental effects of overflows, and wastewater fee impacts and soliciting public comment on the proposed LTCP. Using decentralized controls can expand the public involvement component of compliance efforts.

The small-scale, distributed, and often vegetated nature of decentralized controls differs greatly from the largely unseen conventional stormwater and wastewater infrastructure network. Adopting decentralized control practices educates and connects the public to water management and treatment problems. Bringing stormwater management into neighborhoods, right-of-ways, and onto individual lots moves overflow solutions from an unseen and abstract notion to a scale that engages the public. These efforts allow individuals to see first-hand the connection between rainfall and runoff. The positive response that often accompanies the installation of decentralized controls engages the public and provides greater opportunity for them to be involved in the process of determining appropriate methods of control.

Figure 2-2. Stormwater Management at the Oregon Convention Center is as Much an Amenity as an Innovative Control System.
Photo reprinted with permission from Mayer/Reed.

2.5 Alternative Management of Stormwater

2.5.1 Comparison of End-of-Pipe vs. Decentralized Controls

The old model of excellent drainage through a pipe network is called into question with the capacity and treatment concerns associated with an end-of-pipe system. These systems have a net effect of creating a sustained peak flow period which, while not exceeding predevelopment storm flow peaks in volume at a given point in time, extends the time that peak volume flows. This has unintended consequences downstream, such as increased erosion.

Although the initial impetus for applying decentralized controls is water quality, a number of other planning goals can be met, saving money, energy, and time. Rather than whisk the water away to a detention pond, a deep pipe, or conveyed to a treatment facility that is far away and out of site, decentralized controls allow stormwater to be visibly and attractively integrated into the development pattern (Figures 2-3 and 2-4). Improved stormwater quality goals are met and other planning goals such as provision of recreation amenities, creation of habitat areas, or establishment of flood control buffer zones can be met with the same land use strategy.

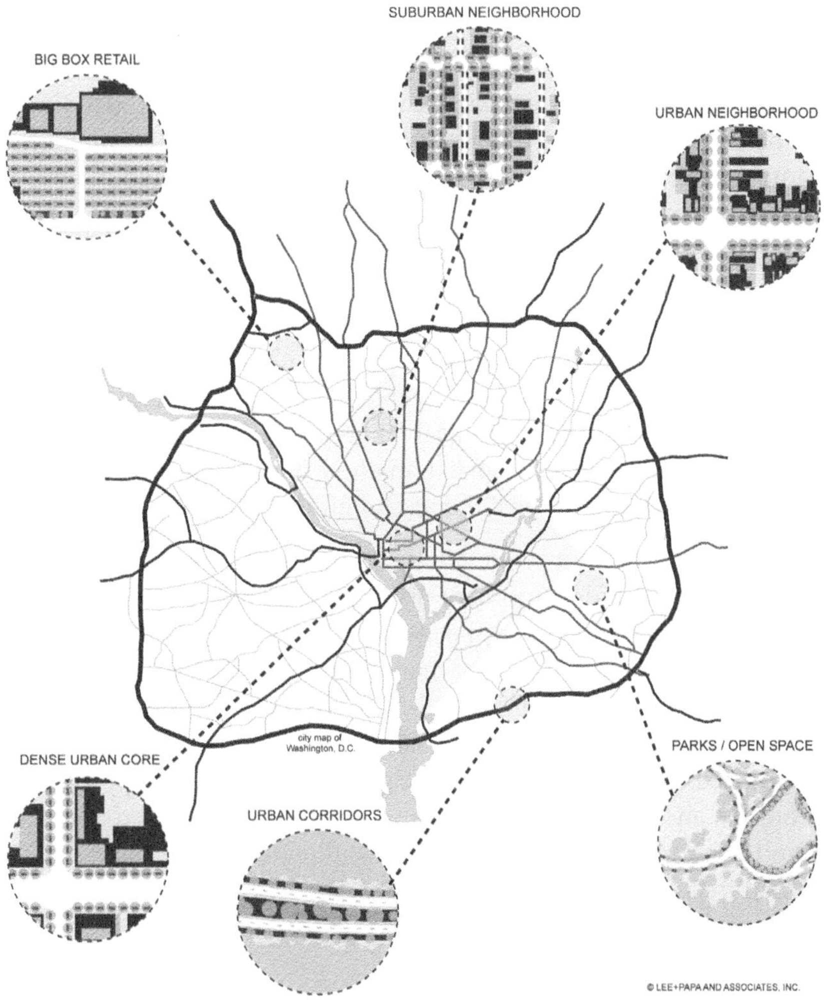

Figure 2-3. Decentralized Control Application for Urban Land Uses.

Figure 2-4. Concept of Stormwater Management at the Portland (Oregon) Convention Center.
Graphic reprinted with permission from Mayer/Reed.

Decentralized stormwater controls use a treatment train approach that treats each part of the system, no matter how small, as an opportunity to improve water quality and reduce runoff volumes. For example, infiltration and treatment of stormwater through the biofiltering processes of the roots in a bioretention cell results in cleaner stormwater and lengthens the time of concentration for any given rain event, having the effect of flattening the magnitude of the peak flow. Disconnection of downspouts from combined sewers reduces the volume of water flowing into the pipe system. Redirection of flows from disconnected downspouts into grassy swales or rain gardens slows the water close to the source, which both reduces the erosive power of the water and provides more time for infiltration, evapotranspiration, and later entry into any storm sewer pipe system. Some sites will lend themselves to the development of a recirculating system of stormwater between ponds, creating a visual amenity while treating the water and keeping it out of the larger stormwater system (Perello and DePalma, 2006).

Decentralized controls can improve conditions both as retrofit applications on existing development and as part of infill development projects. Urban infill development will likely, in the absence of a stormwater management plan at the site level, add to the volume and water quality problems in a community. Given that many retrofit opportunities exist on small parcels (<1 acre), they may fall outside of the normal "limits" to growth" approach of regulating runoff on existing stormwater infrastructure. Rather than place a moratorium on infill development, development can proceed in conjunction with the use of a distributed approach that focuses on performance of controls and impact on the existing systems of development. As part of the overall planning and implementation process, the installed decentralized controls should be mapped and recorded for future monitoring. Property owners should be advised of the stormwater function of the control features and a regulatory process to oversee requests for modifications over time should be put into place.

2.5.2 Appropriate Locations for Use

Several development concepts for urbanizing areas have emerged under a wide range of titles in the past decade that promote themselves as "greener" solutions for development. These include Sustainable Development, New Urbanism, Leadership in Energy and Environmental Design (LEED®) standards: Traditional Neighborhood Development, Smart Growth, and Green Growth. All of these approaches are working towards a vision of development that provides human living conditions that can accommodate denser populations while minimizing negative impacts on the environment. These approaches largely focus on the attributes of a building envelope – materials from environmentally friendly or sustainable sources, density trade-offs that leave some parcels open in exchange for greater building density elsewhere, and building alignments which are promoted as improving communication at the pedestrian level. Implicit in these approaches is that by following the approach advocated, healthier communities will emerge. All of these approaches are promoted as having lower impacts from development on the environment but decentralized controls represent an approach which is the only one that is hydrologically based. Figure 2-5 illustrates the effects that decentralized controls can have on the urban hydrologic cycle.

A number of new development proposals have set themselves to conform to the LEED standards that are promoted by the National Building Council. Most of the points available to be earned are for building practices that are not directly stated for stormwater management. Very few points are specific to stormwater but many decentralized controls are valued. An example of this is a green roof; it will not earn stormwater points but it will earn Sustainable Sites credit for new construction as contributing to reducing the heat-island effect. In a similar vein, using decentralized controls such as rain barrels, cisterns, rain gardens, or infiltration trenches will not earn credit for stormwater, they will earn Water Efficiency LEED points. Other practices will not earn additional points, but they are mandatory measures; for example, installing decentralized controls allow for less site disturbance than would occur if a conventional centralized stormwater control system was put into place (U.S. Green Building Council, 2005). The potential for credits for a retrofit project using decentralized controls, based on performance, spans several LEED categories, regardless of project type.

Although they focus on development patterns and strategies, most of the aforementioned approaches do not sufficiently address stormwater control. A decentralized control approach can be used in all green approaches to community development to improve their stormwater control results. Decentralized controls are applicable to each of these development approaches because stormwater flows through all types of development. Specific BMPs or control approaches may or may not be applicable depending onsite conditions, but the strategy of decentralized stormwater controls is feasible in nearly every situation.

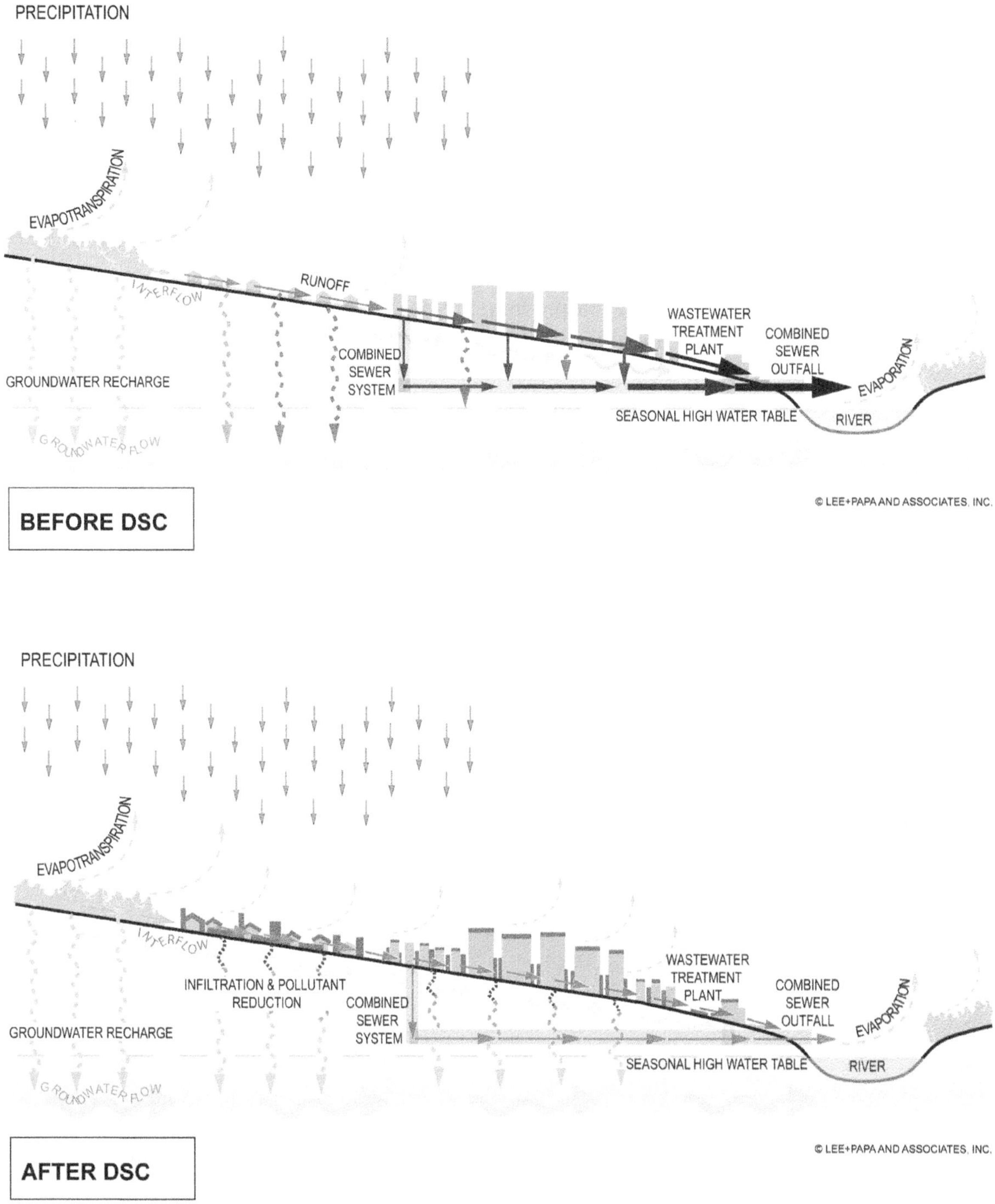

Figure 2-5. Decentralized Controls (DSC) Influence on the Urban Hydrologic Cycle.

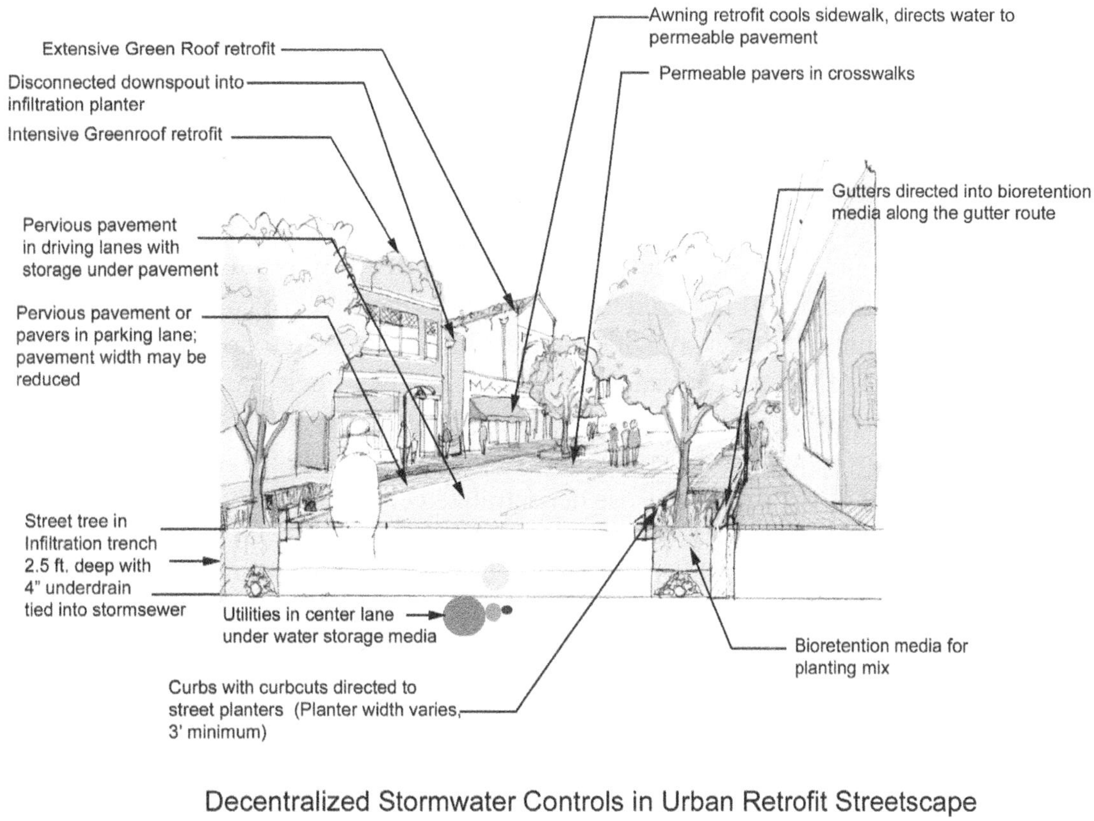

Figure 2-6. Streetscape Design with Fully Integrated Decentralized Controls.

Several of the popular green practices are site- or parcel-specific rather than watershed specific, due to both political and economic reasons of scale and location (Figure 2-6). However, stormwater management is increasingly being mandated to be viewed from a watershed level by U.S. EPA and they promote the use of watershed planning in their approach to state and local governmental regulations. The efficacy of the decentralized stormwater controls is supported by data (as shown in the WERF Phase I document) as a means of lowering impacts from development on the environment more than conventional stormwater practices. Decentralized controls should be part of every development strategy, regardless of infill or newly urbanizing environment because they are effective, attractive, and sustainable tools for addressing stormwater problems. Decentralized controls are site-specific tools that are applied to individual sites as part of a system of controls designed to meet larger scale stormwater management goals.

2.5.3 Need for Policy and Ordinance Changes

In some cases, the use of decentralized controls will be precluded by the ordinances that local governments have adopted over the years in the name of managed development. Twenty years ago, when the first stormwater management plans were widely developed, the end-of-pipe, stormwater detention pond was thought to be the most efficient solution to the problems of stormwater from urbanizing development. Over time, this has proven to be a strategy that has been imperfect at best and better ways to address stormwater based on the local hydrology and watershed and sewershed drainage patterns have been researched and shown to be superior in the areas of volume reduction and water quality (Derry, et al., 2004). Changes in the codes and ordinances have not caught up with current best management practices in many cases but there is movement across the United States in a direction of addressing stormwater through more sustainable solutions. A number of codes (several are highlighted below) could prevent the use of decentralized controls (NRDC, 2002).

- Do building and construction codes restrict core functions of decentralized controls (e.g., infiltration, stormwater storage)?
- Are there adopted standards and specification for decentralized control practices?
- Is rooftop or site runoff required to be directed to the street or collection system?
- Are there minimum driveway widths or restrictions on shared driveways?
- Is there a prohibition on permeable pavements or "two-track" lane designs?
- Do standards prevent the temporary ponding of surface runoff?
- Do regulations require curbs and gutters or prevent open drainage systems?
- Are there incentives in place to use decentralized controls?

Greenville, North Carolina; Chicago; Stafford County, Virginia; and Pennsylvania are a few of the places that have amended their subdivision and zoning ordinances to allow decentralized controls to be used to meet their stormwater requirements. Stafford County amended the local development codes, waiving previous requirements like curbs, gutters, and sidewalks and facilitating the use of other distributed practices. The county revised its stormwater manual to describe distributed practices and how to incorporate them into site design (U.S. EPA, 2006). The Commonwealth of Pennsylvania recently developed and published model codes that encourage decentralized applications (Brown, 2007). Developers must conform to the practices in local ordinances; as more ordinances reflect best practices, more sustainable solutions will be implemented.

A management system with decentralized controls is also able to respond to political and technological changes, whether it is the public's desire for more urban green space or improvements and advances in BMP applications. As a distributed approach, it is inherently flexible and adaptive; monitoring of the system can allow for adaptive management by responding to measured performance outcomes and modifying the program to satisfy new requirements or conditions. Incremental funding and implementation does not tie a municipality to one prescriptive approach, allowing for evolution to occur as a program matures. This aspect of decentralized controls is well suited for the complex and dynamic urban environmental environment.

CHAPTER 3.0

PLANNING STRATEGIES FOR RETROFITS

3.1 Introduction

Planners and stormwater managers must formulate retrofit strategies that consider the various levels of executive governance, including elected officials and career service employees. Municipalities must be able to meet the regulatory mandates of the larger governing bodies such as U.S EPA and state environmental agencies as they grapple with practical solutions for urban retrofits and CSO reductions. Decentralized stormwater controls are a set of site-specific tools that may be employed strategically at the planning level in the form of green infrastructure planning and at the enforcement level at the permitting and monitoring stages.

In this chapter, the researchers show how to incorporate a decentralized control approach into the urban planning process for urban retrofits and mitigate the problems posed by CSOs. The information presented here will benefit the planners and decision makers who, along with engineers, play an important role in a decentralized management approach. It is critical that technical and planning staffs and elected officials coordinate their efforts to create distributed, yet integrated system.

Decentralized systems are adaptable to both micro- and macro-scale applications to solving water quality problems. The systems meet stormwater management goals and planning and design needs for urban settings to the benefit of urban dwellers and downstream recipients of stormwater. Controlling urban runoff with a decentralized control approach provides communities with a creative solution that mitigates the identified stormwater problems. The system adds value to the community in the form of a variety of treatment applications that are attractive and cost less to implement and maintain when considered in a life-cycle cost context. This process allows a municipality to meet its regulatory obligations while enhancing its natural resources.

Decentralized controls are suited for urban retrofit situations because they reduce the volume and improve the quality of stormwater from the built environment. Decentralized controls offer a strategy to achieve runoff management targets by improving conventional infrastructure performance. Most decentralized controls have a small footprint when built, so the scale and effectiveness of these BMPs makes them particularly well suited to the often land-locked and land-limited urban and ultra-urban situations. They are also flexible in how they are organized in an application and can be linked in a number of ways to be used on site level projects or in watershed scale applications.

Engineered conveyance and end-of-pipe enhancements have been widely implemented and codified in the United States. However, these solutions have not yielded the improvements to water quality and reductions in water volume that are needed to solve the pollution problems associated with MS4 discharges and CSOs and other alternative solutions are needed. In both practice and research, decentralized stormwater controls have been established as a viable alternative strategy for improving water quality and achieving CSO reductions. As implied by its name and explained at length in the Phase I report, decentralized controls integrate an interconnected system of innovative BMPs and practices into the more commonly used centralized methods of stormwater and wastewater management. The Phase I report investigated the technical issues surrounding the application of decentralized controls. Greater adoption of these measures, however, raises not only technical considerations, but also planning, regulatory, and economic ones as well.

Integrating decentralized controls into an urban environment is accomplished by using structural and non structural tools. Examples of non-structural tools are included in Table 3-1 and a planning process approach is shown in Figure 3-1.

Table 3-1. Non-Structural Tools for Water Quality Protection.

Non-Structural Tools	Function
Comprehensive master planning	▪ Provides a coordinated approach with consideration of water quality protection.
Stream and wetland setbacks	▪ Place limits on development activities within certain distances of streams and wetlands. ▪ Maintain natural flood and erosion control functions. ▪ Protect infrastructure and homes from moving streams. ▪ Use protection zones. o Based on public health and safety services. o Establishes minimum setback widths. o Details permitted and prohibited structures and uses. o Allows for non-conforming structures and uses. o Includes variance section to maintain buildability and provide flexibility.
Conservation design	▪ Allows for partial conservation of natural resources while development occurs. ▪ Allows homes to be closer together while remainder of development area is in open space. ▪ Requires understanding of site's soils, topography and natural features.
Distributed stormwater management	▪ Stormwater controls are site features implemented at the source of impacts. ▪ Potentially reduces stormwater infrastructure costs. ▪ Promotes public participation and public education in pollution prevention and maintenance stormwater management practices. ▪ Satisfies National Pollutant Discharge Elimination System (NPDES) Phase II stormwater requirements. ▪ Application for retrofit projects.

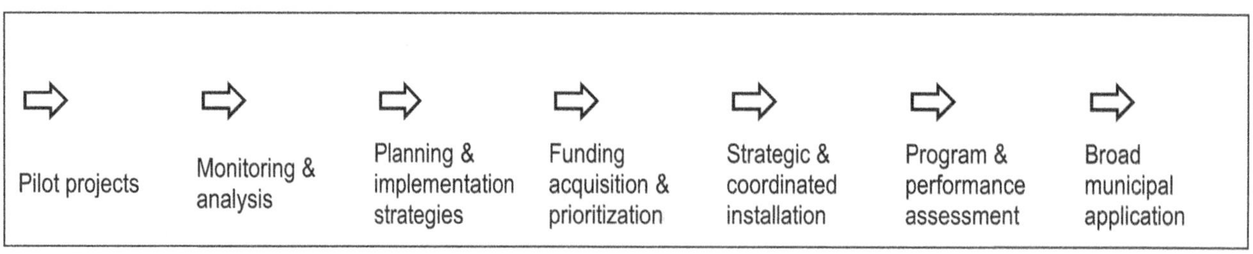

Figure 3-1. Planning and Implementation Process.

3.2 Integrated Planning and Design

Traditionally, stormwater management has been part of a compartmentalized system in cities that separates planning and operations into separate divisions. At the local level, departments of public works, water, or transportation have assumed oversight and responsibility for stormwater management. These departments are charged with enforcing local regulations and interacting with regulatory agencies at the local, state, regional, and federal level. Local planning departments have had oversight and responsibility for technical advisement of other community development and quality of life issues such as housing density allocations, economic development, and land use allocation. Planning departments typically function as umbrella organizations at the local level and are charged with producing recommendations for action that are then given to a political body for decision making and funding decisions. These departments are also often responsible for seeking funding for community development projects that relate to planning issues. Planning agencies typically interact with public works bodies as technical advisors and on broader governmental agencies for planning. Stormwater is addressed at several levels of governmental operations; the result is a "nested" system of permits and jurisdictions reflecting the nature of multi-scaled efforts targeting water quality and water volume. The varying responsibilities within local government result in contrasting agency structures with planning set up in an advisory capacity and public works set up in an enforcement capacity. Local governments may or may not have a bureaucratic structure to encourage coordination between departments to solve problems that impact environmental quality in the environment.

> "Water resource planning spans such a variety of plan types and scales of implementation, that specific guidelines on level of detail cannot be given....water resource planning has long implementation horizon (decades); therefore need adaptive management which views management policies as adaptive and flexible and experimental rather that fixed policies" (Philadelphia Water Department) (Maimone et al., cited in Novotny and Brown, 2007, pp. 302–304).

The elected body of officials responsible for settling on local budgets is the third leg of governance. Stormwater retrofits can only happen effectively when the political body and the enforcement body of the government (typically public works) fund the plans developed by the planning body of the government.

In addition to the local volume of stormwater generation and need for reduction, stormwater has an additive effect on downstream communities. Upstream development can create silt and trash problems that can overwhelm a watershed at the downstream point. Thus, it is best for stormwater managers to find solutions by developing a cross-jurisdictional approach. Overlapping jurisdictions, the diffuse nature of the inputs into the final stormwater or CSO problem, and the broad nature of the effects of stormwater on a community warrant an integrated planning effort. Decentralized stormwater controls are part of a holistic and integrated ecosystem based problem solving approach that spans the jurisdictions that govern the sources of the water (Newman and Kenworthy, 1999). U.S. EPA has advocated a watershed-based planning model because of the overlapping political issues that are best reconcilable at the watershed level. The Philadelphia Water Department has determined that meeting the water quality targets

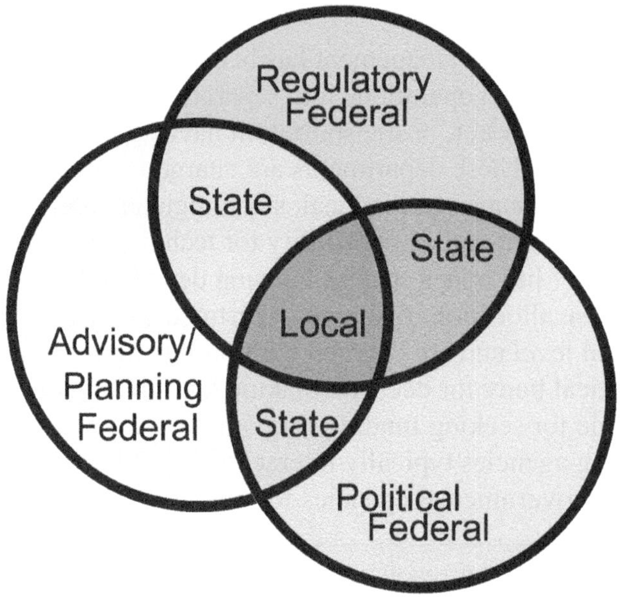

Figure 3-2. Interwoven Influences on Stormwater and CSO Policy and Implementation.

100% of the time is not possible and has developed an alternative strategy that accounts for and allows for incremental progress to be attempted, monitored and valued (Mamione et al., cited in Novotny and Brown, 2007). This approach uses a range of tools: regulatory, structural, coordination/incremental implementation, scheduling, clarity of jurisdictional responsibility in plans, and appropriate levels of detail for planning types. Three levels of water quality efforts are measured:

1. Meeting water quality goals in dry weather;

2. Promoting practices that create fishable conditions at all times; and

3. Meeting stormwater and human health water quality goals during wet weather.

Each level is progressively more difficult to attain, but each level provides measurable and meaningful improvements to water quality.

The Philadelphia example highlights an important point. Applying decentralized controls, or any stormwater control techniques, across an entire watershed or jurisdiction is daunting, from both an implementation and financing perspective. As a decentralized approach to stormwater management, decentralized controls offer the opportunity for incremental and strategic application of BMPs. Decentralized BMPs can be targeted in a neighborhood that has immediate collection system problems, like sewer backups or street flooding. Likewise, decentralized controls can be applied in watershed locations that can positively alter the hydrologic influences and timing on the watershed or the hydraulic impacts on a collection system. This flexibility creates multiple layers of planning that can produce many alternatives for implementation. This allows for immediate attention to be given to localized problems as well as the additive effect of multiple targeted projects over time.

To succeed with a larger-scale approach (e.g., at watershed and local levels), managers must often develop new procedures to formalize the process of interdepartmental cooperation and to define responsibilities for the stormwater management. At a watershed scale, state or local officials may need to pass legislation to allow for such intergovernmental work to occur. For example, the Philadelphia Water Department calls the effort Adaptive Management, describing the water resource planning approach as one characterized by flexibility and adaptable and open to the input of information from new developments from science or other disciplines (Mamione, 2007, cited in Novotny and Brown, 2007). This is necessary because of the extremely long planning horizon in this field. In some states, such as Idaho, there are stormwater management directives from the federal permitting agency (U.S. EPA), the state Department of Transportation for state roads, and then local jurisdictions for their local ordinances (Idaho Department of Environmental Quality, 2007). Any local effort at changing regulations to allow decentralized stormwater controls in the effort to reduce CSO volumes and for urban retrofit will need to be coordinated with the appropriate larger scale agency with jurisdiction over the water resource.

A regional planning approach is extremely compatible with watershed planning approaches. Jurisdictional structures differ among the states, so this will have varying degrees of ease to bring into a reality. In Duluth, Minnesota (a state with strong regional planning authority) they have integrated their local approach to stormwater into the regional approach (Duluth Streams, 2007). Duluth identifies and describes its approach with the following.

Duluth is part of the NPDES Phase II Stormwater Permitting Process. In March 2003 the City submitted a permit application to the state and began operating under a stormwater permit. The purpose of the stormwater permit is to encourage cities to take an active role in preventing pollution from stormwater. Some recent estimates are that as much as 50% of the pollutants entering our oceans, streams, and lakes are from urban pollution from individuals and their automobiles. Since all stormwater in Duluth ultimately ends in Lake Superior, all citizens have an added responsibility to protect this pristine water and in doing so protect the source of our drinking water.

Because of the Phase 2 rule, there is increased pressure on state and regional planners to formulate policies and strategies to mitigate the existing stormwater situations; CSO mitigation is now a formal part of the planning and implementation. State and federal law adds the enforcement power to planning recommendations but the staffing and daily enforcement component will fall in the realm of the local municipalities charged with reducing CSO events or meeting MS4 permits. Reconciling mandates from outside of jurisdictions with the realities of local budgets falls beyond the purview of this document but the priority of stormwater improvements or CSO reduction targets in the form of adopted plans and budget dollars is an important part of meeting those goals.

The effective use of decentralized controls requires a process for developing a distributed but integrated approach to stormwater management within a municipal jurisdiction. The success of an urban retrofit or CSO mitigation process requires a coordinated effort. The process of transferring stormwater control from a single departmental/engineering/drainage issue is an exercise in "bringing the right groups together." Planners will need to model the positive economic impacts of using decentralized controls; politicians will need to have the political will to pursue stormwater goals that are set for the community through the planning process; and the public works department will need to be supported in its enforcement of construction and compliance deadlines by both political and financial means.

3.2.1 Drivers for CSO Reductions, Urban Retrofits, and Water Quality Improvements

Large-scale direct drivers for specific CSO reductions are set at the federal, regional, and state levels. Municipal stormwater mitigation efforts are often influenced by a number of regulatory requirements, including EPA's Phase 1 and 2 regulations, water quality targets set by watershed-based agencies at the regional level and at state natural resources departments, and LTCPs submitted for U.S EPA approval. Other indirect drivers that will affect goal setting are specific state or local planning targets that require decentralized controls to be part of a development or redevelopment plan. Politically enacting planning targets set by technical planning staffs are subject to a variety of public pressures. Community-based groups, local planning growth management initiatives, the current severity and urgency of the stormwater or CSO problem, and the cost to the community to address the issues will drive both policy decisions and budget considerations for urban stormwater retrofits. Each local jurisdiction will need to determine which drivers are most pertinent to include in their efforts for using decentralized control retrofits to meet the stormwater management goals that are set.

The impetus for improved stormwater management may come from the areas of public health, economic redevelopment, and phased infrastructure improvements. Stormwater runoff and CSOs have been proven to foul the environment and cause public health problems and are an unattractive visual deterrent within a community. The desire to create a healthy, livable, and prosperous municipality is consistent with most political bodies at the municipal level. The CSO reduction and urban retrofit strategies adopted by a municipality can be, in part, driven by these considerations along with the compliance requirements imposed on a community from the outside.

3.2.2 Coordination Between Technical Planning and Political Jurisdictions

Runoff volumes that contribute to MS4s or combined sewer systems may start higher in the watershed, outside of the direct political jurisdiction of a sewer or urban governmental authority. In this case, the larger watershed planning jurisdiction (which may contain multiple urban areas that end up having a cumulative effect within a watershed) is a driver in how to address the external sources contributing to the stormwater problem. In these instances, interdepartmental coordination networks can be established to allow for the sharing of information and for teaming across departments to promote integrated planning efforts from the regional level down to the local level. Local jurisdictions should typically take on the site level identification tasks of identifying the sources that contribute to local stormwater problems or CSOs. Representatives from the various identified groups should be involved in the process of identifying urban retrofit and CSO mitigation drivers (e.g., federal, state, and local regulations aimed at improving water quality). For example, the Milwaukee Metropolitan Sewerage District (MMSD; http://v3.mmsd.com/AboutUs.aspx) manages wastewater from 28 municipalities in a 420-square mile service area. The 28 separately owned and operated sewer systems drain into

300 miles of regional sewers owned by MMSD. MMSD's regional system inherits and is ultimately responsible for stormwater control problems from its municipal contributors.

The process of linking between jurisdictions is both technical and political. In some states, enabling legislation may be required to facilitate coordinated planning across jurisdictions (e.g., Pennsylvania) while other states with strong regional planning traditions (e.g., Minnesota) will not require this step. Reston, Virginia, has developed such a model for coordinated planning across jurisdictions (Brown et al., 2002). In many cases municipalities have to contend with externally generated stormwater problems even while only having control over stormwater problems generated within the municipality.

In California, the Bay Area Stormwater Management Agencies Association (BASMAA) has been working on the problem of linking planning and application across jurisdictions within the San Francisco Bay Watershed. BASMAA represents more than 90 agencies, including 79 cities and six counties and the bulk of the watershed immediately surrounding San Francisco Bay (BASMAA, 2006). They produced technical guidance [*Start at the Source: Design Guidance Manual for Stormwater Quality Protection* (Richman et al., 1999), followed by a companion manual *Using Site Design Techniques to Meet Development Standards for Stormwater Quality*, published in 2003] for distribution with the membership and beyond and also established the structure through which inter-jurisdictional cooperation can be facilitated. The regional authority acts as a coordinating body for each of the entities that participate in BASMAA (2003).

3.3 Governmental Cooperation

In Puget Sound in Washington, a "trickle down" approach to intergovernmental partnerships has been employed to address water quality issues. Since being formed in 1996, the Puget Sound Partnership defines, coordinates, and implements Washington's environmental agenda for the Puget Sound watershed – an area that includes 12 counties, 115 cities, and the lands of 17 tribes. The 17-member partnership includes directors from 10 state agencies, representatives from three federal agencies, one representative of tribal governments, two representatives of local governments (city and county), and a chairperson appointed by the governor (U.S EPA, 2006).

A good example of how a cooperative model of planning can be established at a more regional/local level is modeled in a document created by the Chesapeake Executive Council in 2001. The directive was designed to "address stormwater management to control nutrient, sediment, and chemical contaminant runoff from State, Federal, and District (Washington, D.C.) owned land" (Chesapeake Executive Council, 2001). This Council took a watershed approach to the problem of stormwater runoff and established that there was a mutual benefit to all of the parties involved to control runoff entering the Chesapeake Bay. Broad goals were set that included an inventory of targeted public lands; a number of demonstration projects that would illustrate a variety of infiltration and decentralized stormwater controls such as source control of pollutants, green roofs, rain gardens, and pervious paving; and a commitment to monitoring those projects and disseminating the results promptly. This directive addressed that there was a need to coordinate and encourage information sharing of innovative technologies that are available between the various groups. In this case, it included the Urban Stormwater Workgroup, the Transportation Workgroup of the Land, Growth, and Stewardship Subcommittee, the Federal

Highway Administration, and the State Department of Transportation. This effort produced information that would then trickled down to the local public works departments and other road building entities. The presence of the council provides an umbrella organization that can mediate issues that arise between smaller jurisdictional units.

The Chesapeake Executive Council Directive (2001) set targets for what types of projects could be done and in what type of settings and set target dates and recipients for implementation, monitoring, and reporting. In addition, the directive called for economic analyses of the various projects, education efforts, and identified the organizations that would be accountable to carry out the tasks described. The final three directives in their report are aimed at support for an ongoing discovery and implementation process of improved management:

1. develop innovative stormwater technologies;
2. coordinate with communities and local governments; and
3. measure progress.

Each urban setting will have its particular mix of sources that impact the stormwater runoff. It is the responsibility of the local government to determine which factors are most important for a given place using the resources available from broader scale organizations. Local performance measures and site conditions will determine which decentralized controls are best suited to the stormwater mitigation plan developed for that jurisdiction.

For example, in Massachusetts, two state level organizations, the Massachusetts Office of Coastal Zone Management and the Department of Environmental Protection, work to include representatives of major sources of non-point source (NPS) pollution, industry groups, non-government organizations, homeowners, and the public throughout their planning process. This allows for developing a "program to protect coastal waters without unduly burdening those that are affected by the Coastal Non-point Source Pollution Control requirements." Coastal Zone Management (1994) used the federal guidance document for NPS pollution categories and determined that all existed in Massachusetts.

- urban sources
- marinas/boats
- agricultural sources
- forestry
- hydromodification
- wetlands

The Massachusetts example illustrates that different localities will identify specific stormwater issues and will need to select decentralized stormwater controls that are appropriate to that locality's situation. The design criteria for calculating stormwater runoff are either volume-based or area-based. Each may determine their stormwater design criteria that best fits the locality. Performance standards for either method of determining stormwater runoff are established based on the impact on stormwater reduction goals. The MS4 permit or the LTCP should be used to establish target reductions over a phased in period of time.

3.4 Implementation Alternatives

Each separate stormwater or combined sewer system has unique operational and system characteristics that preclude prescriptive guidance. A few planning and implementation scenarios are the most likely to be used for decentralized control installation and selection. Four likely approaches to encouraging decentralized control inclusion are:

1. Pilot installations by an agency or organization at any level;
2. Watershed-based approaches, which will typically be overseen by a regional-level authority (or higher);
3. Building and site redevelopment projects that work to implement distributed controls as part of their building process; and
4. Large-scale urban revitalization projects that will often be a collaborative process between a number of agencies at a variety of levels, including private developers.

For each municipality, the particular stormwater drivers will need to be identified. This will lead to the next step of determining which controls are appropriate to address the problem. The process of BMP identification will need to be undertaken parallel to the processes of determining funding sources for implementation and monitoring protocols for oversight of the final control system.

3.4.1 Pilot Installations

An agency or municipality has a great deal of flexibility in using a pilot installation to demonstrate the efficacy of decentralized controls on an urban retrofit site. First, there are many opportunities in an urban context to insert decentralized controls or a treatment train of distributed practices on a small scale and then monitor the results. If the locality would treat the decentralized method as a non-conforming stormwater approach, then the appropriate approval process can be followed to acquire permission to implement a pilot project. Pilot programs may be framed in the context of other community development goals such as traffic calming, recreational amenity creation, habitat and wildlife corridor preservation and creation, pedestrian/walkability initiatives, public health improvement, environmental education, or property value enhancement. A pilot program can also be undertaken as a stand-alone project, such as retrofitting permeable pavers into a parking lot to reduce runoff volumes. Figure 3-3 illustrates a green alleyways program in Vancouver, British Columbia, aimed at reducing runoff from traditional impervious alleys.

Figure 3-3. Permeable Pavers and Porous Surfaces were used to Retrofit a Traditional Alleyway and Reduce Stormwater Runoff in Vancouver, Canada. Photo courtesy of City of Vancouver Greenways Program.

The City of Vancouver has also undertaken an innovative greenways program aimed at making transportation corridors more pedestrian friendly while also integrating decentralized stormwater controls. As part of the effort, the city created traffic calming and improved pedestrian line-of-sight at intersections by extending curbs into the street. The city's Streets Department has begun to fill the curb extensions with permeable soils and vegetation as a decentralized management technique (Figure 3-4). Stormwater is directed into the extensions and any excess flow passes into the collection system. Because the stormwater retrofits are installed in conjunction with the roadway improvements, the only additional costs are for modifications to curb construction and additional soil excavation. The additional costs are offset by up to $4,000 of savings that would have been required for conventional stormwater infrastructure realignment originally proposed for the project. In addition, a portion of the systems are maintained by volunteers under a program in which local residents adopt and care for a portion of city green space [Natural Resources Defense Council (NRDC), 2006]. This program highlights the integrated, cost effective, and creative manner in which decentralized controls can be retrofit into the urban environment.

Figure 3-4. Vegetated Curb Extensions Provide Cost Effective Decentralized Stormwater Management in Vancouver, BC. Photo courtesy of City of Vancouver Greenways Program.

From 1983-1995, the Tokyo Metropolitan Government undertook an Experimental Sewer System project that has proven to be highly effective in reducing runoff volumes and increasing groundwater recharge. The city used a system that included 5.5 square miles of infiltration pits, 122 acres of permeable pavement, and more than 175 miles of infiltration trenches. The results of this project were described thusly (Becker and Raasch, 2002).

It was determined that the cost of using micro-scale infiltration techniques such as permeable pavement and infiltration inlets cost approximately 33% less than conventional open pond detention systems and 10% of the cost of storage vaults. This alternative was so much more cost effective due to the high cost of land and complexity of the existing infrastructure. Perhaps equally impressive was the reduction in monitored storm drain flow volumes of up to 50%. Most importantly, was the reduction in CSO events from 36 to 7.

The Japanese system included infiltration in their planning process and established a political base for the initiative; the demonstration project has spawned several other pilot projects in other Japanese cities. The results, including volume reduction, maintenance, and public acceptance are continuously being monitored (U.S. House of Representatives, 2003).

3.4.2 Watershed-based Approaches

Watershed-based approaches have two central challenges: either the watershed is multi-jurisdictional or the jurisdiction has multiple watersheds. Water flows through a range of governmental entities with a variety of regulations and regulatory procedures. Coordination of the entities involved to achieve consistent retrofit action throughout a watershed is an organizational challenge. It is the logical approach to take based on science and water movement, but it is one of the more challenging approaches to take in today's regulatory environment. BASMAA in California is an example of an approach that encompasses a multitude of levels of government and regulation in a comprehensive and cooperative effort at the watershed level to work towards achieving stormwater management goals.

The German experience with a watershed-based approach to stormwater and CSO mitigation was pioneered in the Emscher catchment. Micro-scale approaches have been used in the Emscher catchment as a means to control runoff volumes and reduce flooding. Education informing the public of the need to change from an efficient drainage system that was well established but which also caused flooding was a central feature of the effort. The Emscher catchment area encompasses 330 square miles, 2.5 million people, and 17 cities; over 98% of the infrastructure uses CSO systems with approximately 60% of the land area being impervious. The result of building and disconnection efforts is estimated to have reduced volumes in the peak of the two-year storm event by 10%. The majority of the work was undertaken by private initiative, which was self reported. The primary driving force of the private initiative was a desire for environmental protection, but this also included a secondary desire for fee reduction and financial incentive realization (U.S. House of Representatives, 2003).

3.4.3 Building and Site Redevelopment and Urban Retrofit

At the building and site redevelopment scale, managers select decentralized controls on the basis of how the BMPs and practices fit into the overall treatment train and by what the specific site will allow, including area of ground available, infiltration capabilities, relation to other off-site treatment and conveyance mechanisms, and local ordinances. The same scenario applies to urban retrofits. The design for BMP placement is specific to the site, and ordinances must be structured to allow for site-specific flexibility while mandating performance standards for stormwater management.

A performance zoning concept known as the Green Area Ratio (GAR) has been piloted in Berlin, Germany (Figure 3-5). The GAR is a locally controlled, ratio-based rating system that values site development proposals before permitting. Thresholds for minimum GAR scores are required depending on the land use zoning. The GAR approach is presented as a practical and flexible approach to environmental quality improvement. There are three guiding principles of the GAR (Fitzgerald, undated).

1. Modest, decentralized green infrastructure can have a cumulative effect;
2. Each parcel must in some way contribute to the city's green infrastructure; and
3. Environmental impacts of development must be mitigated on site.

Weighting factor / per m² of surface type	Description of surface types
Sealed surfaces **0.0**	Surface is impermeable to air and water and has no plant growth (e.g., concrete, asphalt, slabs with a solid subbase)
Rainwater infiltration per m² of roof area **0.2**	Rainwater infiltration for replenishment of groundwater; infiltration over surfaces with existing vegetation
Partially sealed surfaces **0.3**	Surface is permeable to water and air; as a rule, no plant growth (e.g., clinker brick, mosaic paving, slabs with a sand or gravel subbase)
Semi-open surfaces **0.5**	Surface is permeable to water and air; infiltration; plant growth (e.g., gravel with grass coverage, wood-block paving, honeycomb brick with grass)
Surfaces with vegetation, unconnected to soil below **0.5**	Surfaces with vegetation on cellar covers or underground garages with less than 80 cm of soil covering
Surfaces with vegetation, unconnected to soil below **0.7**	Surfaces with vegetation that have no connection to soil below but with more than 80 cm of soil covering
Vertical greenery up to a maximum of 10 m in height **0.5**	Greenery covering walls and outer walls with no windows; the actual height, up to 10 m, is taken into account
Greenery on rooftop **0.7**	Extensive and intensive coverage of rooftop with greenery
Surfaces with vegetation, connected to soil below **1.0**	Vegetation connected to soil below, available for development of flora and fauna

Figure 3-5. Berlin Weighting Factor for Greening Technologies. From www.stadtentwicklung.berlin.de.

The GAR is similar to the Floor Area Ratio (FAR) in that credit is given based on technologies and materials used. However, it is site specific and building permits and occupancy permits will not be issued without a GAR score that meets the zoning requirement. Thus, whereas the FAR provides incentives, which allow greater density within an area, the GAR provides a standard that must be met to allow for any development or redevelopment of a site.

Adopted in January 2007, the Seattle Green Factor is an alternative approach for urban stormwater control similar to the Berlin GAR (Figure 3-6). The Green Factor is a landscaping requirement for commercial areas that stipulates that 30% of a site must be vegetated. This system encourages multiple layers of visible plantings and plantings in the public rights-of-way adjacent to the properties. The system is flexible and weights different landscaping practices according to their effectiveness. The square footage of each practice is multiplied by its green factor and then aggregated with the score of each additional practice to satisfy the requirements. For example, asphalt and concrete have a green factor of 0, permeable pavements 0.6, and green roofs 0.7. Bonuses are also provided for using rain water harvesting and low water-use plants.

This regulatory construct is interesting because it is not stormwater specific, nor does it contain specific discharge performance requirement. However, because of the practices selected for green factors and the benefits gained by adding vegetation and other green infrastructure practices, this policy will beneficially impact the volume of stormwater runoff.

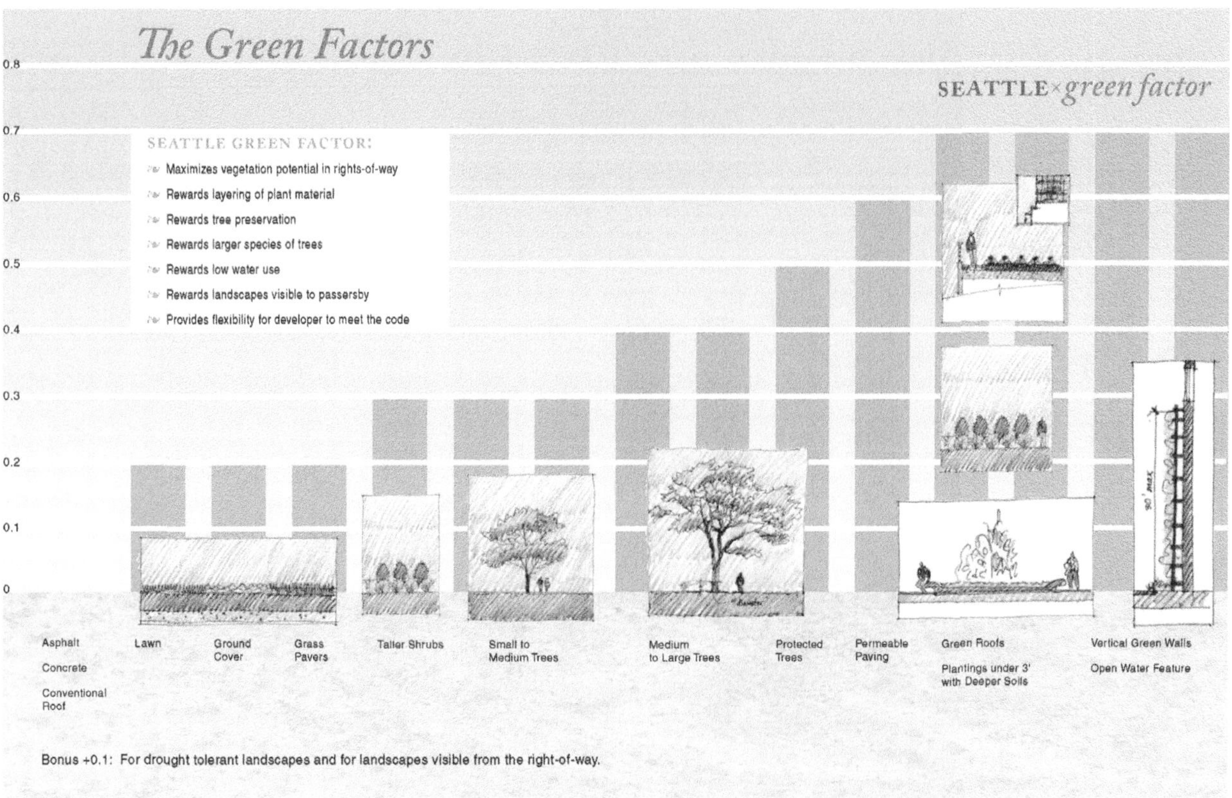

Figure 3-6. Seattle Green Factors. Adapted from Seattle Public Utilities.

To help educate and facilitate the implementation of the green factor into new projects, the city created a worksheet, shown in Figure 3-7, that can be used to determine compliance with the Green Factor requirements.

3.4.4 Large-Scale Urban Revitalization

Urban revitalization projects provide the opportunity for larger areas of distributed urban retrofits to be included in project planning and implementation than might otherwise occur. Figure 3-8 provides an example of decentralized controls that may be employed in dense, urban redevelopment areas. The Anacostia area of Washington, D.C. is an example of a large-scale urban revitalization effort that will include urban retrofits as part of the project planning. Specific environmental standards were adopted that apply to each project within the redevelopment area. The standards encompass four separate, but related, areas:

1. integrated environmental design;
2. green building;
3. stormwater control; and
4. site planning and preservation.

SEATTLE green factor

FINAL VERSION 3-8-07

		enter sq ft of parcel		You need at least 0.300
	Parcel size *(ENTER THIS VALUE FIRST)**	1	SCORE	-

	Types of Area**	Square Feet	Factor	Total
A	**Vegetation planted with a soil depth of less than 24"**			
1	Lawn or grass pavers or ground covers	enter sq ft: 0	0.2	-
2	Plants and shrubs 3' and higher at maturity	enter number of plants: 0 → 0	0.3	-
B	**Vegetation planted with a soil depth of more than 24"**			
1	Lawn, grass pavers or other plants less than 3' tall at maturity	enter sq ft: 0	0.7	-
2	Shrubs taller than 3' at maturity - calculated at 16 sq ft per plant (typically planted no closer than 18" on center)	enter number of plants: 0 → 0	0.3	-
3	Tree canopy for "small trees" in SDOT's Street Tree Planting Schedule or equivalent canopy spread of 15' - calculated at 50 sq ft per tree	enter number of plants: 0 → 0	0.3	-
4	Tree canopy for "small/medium trees" in Street Tree Planting Schedule or equivalent canopy spread of 20' - calculated at 100 sq ft per tree	enter number of plants: 0 → 0	0.3	-
5	Tree canopy for "medium/large trees" in Street Tree Planting Schedule or equivalent canopy spread of 25' - calculated at 150 sq ft per tree	enter number of plants: 0 → 0	0.4	-
6	Tree canopy for "large trees" in in Street Tree Planting Schedule or equivalent canopy spread of 30' - calculated at 200 sq ft per tree	enter number of plants: 0 → 0	0.4	-
7	Tree canopy for preservation of "exceptional trees" or trees with trunk diameter exceeding 24" at four and one half feet above the ground. calculated at 250 sq ft per tree	enter number of plants: 0 → 0	0.5	-
8	Permeable paving that drains only itself. It must be at grade. - calculated per square foot	enter sq ft: 0	0.6	-
C	**Green roofs** - 4" minimum soil depth at time of planting	enter sq ft: 0	0.7	-
D	**Vegetated walls**	enter sq ft: 0	0.7	-
E	**Water features (fountains) or rain gardens** (where allowed by SPU)	enter sq ft: 0	0.7	-
	sub-total of sq ft =	0		
Bonuses				
F	**Landscaping using drought tolerant plants or where at least 50% of annual irrigation needs are met from non-potable sources**	enter sq ft: 0	0.1	-
G	**Landscaping visible to passers-by from adjacent public right of way or public open spaces**	enter sq ft: 0	0.1	-
			green factor numerator =	-

** Do not count public rights of way in parcel size calculation.*

*** To calculate your green factor score, you may count the landscape elements that are in public rights of way if they are contiguous with the parcel.*

Green_Factor_Worksheet Page

Figure 3-7. Seattle Green Factor Worksheet. Adapted from Seattle Public Utilities.

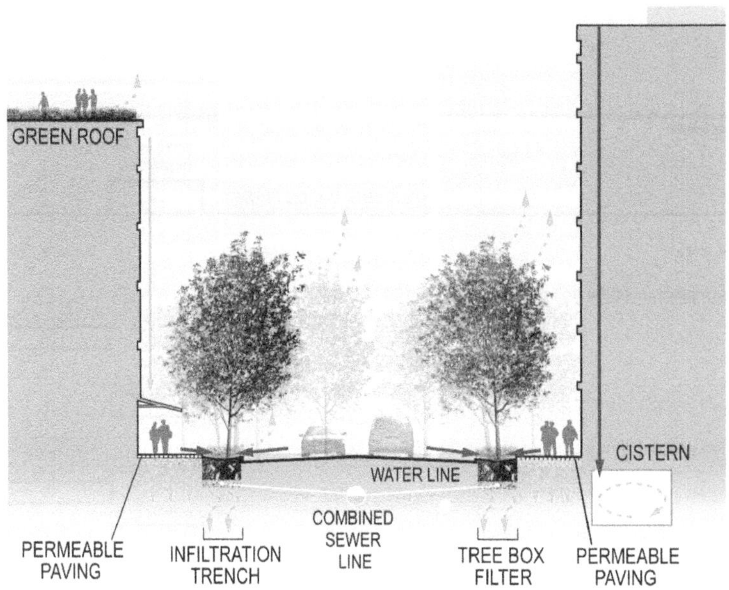

Figure 3-8. Typical Decentralized Control for Dense, Urban Application.

With the redevelopment area along the banks of the river, stormwater and CSO control is an integral part of the standards. The standards require retaining the first inch of runoff onsite and treating for the volumes up to the two-year storm. Importantly, the standards provide a hierarchy of preferred stormwater control options with vegetated, decentralized controls identified as the most preferred option (Anacostia Waterfront Corporation, 2007). The preference for vegetated controls is due largely to their demonstrated performance, but also to their ability to provide other environmental, energy, aesthetic, and economic benefits. These standards have the ability to impact a large-scale, multi-decade redevelopment process and aim to integrate more sustainable control strategies directly into the redevelopment process.

Other cities have similarly implemented more progressive ordinances and provided innovative incentives to facilitate decentralized control adoption for urban redevelopment efforts. Chicago has a multi-faceted program aimed at quickly increasing the use of decentralized controls and providing onsite control of stormwater. City ordinance requires onsite capture of the first half inch of rainfall or a 15% reduction in imperviousness in relation to baseline site conditions. In addition, the velocity of a site's runoff entering the sewer system is governed and restricted by the capacity of the system. To complement the regulatory requirements, the city has implemented a number of incentives for decentralized controls. A green permitting preference that allows for expedited review of projects using decentralized controls has been initiated. The city also provides grants for green roofs and cool roofs. Annually, the city provides twenty $5,000 dollar grants for small-scale commercial and residential green roofs; in 2006, the first year the grants were offered, 126 applications were received for the twenty available grants. Programs are also in place for subsidized rain barrel distribution, often targeted at communities with localized CSO issues and metered pricing has been initiated in the city (Wise, 2007b).

Philadelphia has also instituted a regulatory approach complemented with a series of incentives. City ordinance requires the onsite capture of the first inch of rainfall. To assist developers in meeting this requirement, the city offers pre-design consultation; expedites permits for projects that use green roofs, permeable pavements, and bioretention; and provides green roof tax credit of 25% of construction costs up to $100,000 (Wise, 2007b). The city has also instituted a program that reclaims vacant lots for green infrastructure improvements and decentralized control implementation. The initiative has been an innovative method of incorporating decentralized controls into the city to improve stormwater management, but it has provided economic benefits as well. University of Pennsylvania research has indicated that vacant lot improvements increase surrounding housing values up to 30% and tree plantings result in up to a 10% increase (Wise, 2007a). Efforts from this program benefit private citizens by contributing to a property value increase and public government by increasing tax revenue.

3.5 Implementation and Mapping

Mapping of decentralized control locations as part of implementation is vital to long-term effectiveness. The institution of distributed measures for stormwater management creates an infrastructure system that should be legally documented, especially on large-scale retrofits. Many layers of infrastructure interact with the natural environment, shown in Figure 3-9, and need to be considered when selecting decentralized controls. Stormwater reduction benefits depend on the presence and maintenance of the decentralized controls. If the presence of the decentralized controls is not legally recorded and monitored, the success of a system is jeopardized.

Decentralized control treatment trains can be organized by watershed and drainage patterns. Water does not flow based on property or political boundaries unless the flow is manipulated to do so. Typically, it is not manipulated to do so and the water crosses various boundaries that are defined by a host of jurisdictions from the local level all the way to the federal level of government. Ideally, a planning process will identify the appropriate level of governance to evaluate proposed decentralized control strategies for stormwater management.

Decentralized stormwater controls are an important tool for bridging gaps in the existing green infrastructure that may be identified in the community. They are engineered solutions that enhance and connect existing green spaces; as linking elements between larger green hubs, they may provide valuable habitat corridors as well as other health benefits to a community. The locations of the decentralized controls should be included in a map of green infrastructure in a community and represents another planning resource for determining where to focus efforts. In addition, their locations must be noted on stormwater control maps so that appropriate scheduling of maintenance and monitoring may be done by the appropriate agencies that focus on stormwater management. Decentralized controls are best applied when they consider the relation of the natural environment to the built environment and its associated infrastructure, illustrated in Figure 3-9.

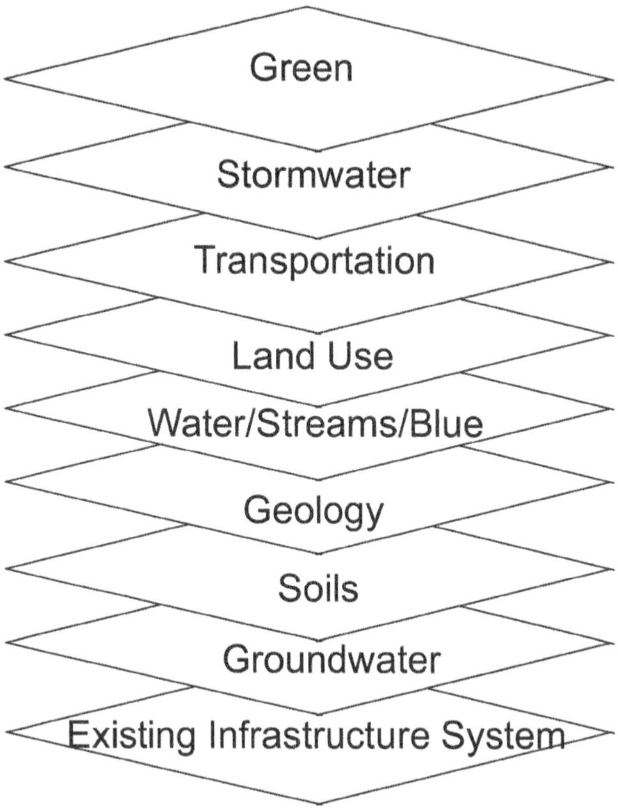

Figure 3-9. Physical Layers of Infrastructure and Natural Resources.

3.6 Land Use Types Suitable for Decentralized Controls

Vegetated landscapes are the most readily converted to an urban retrofit with decentralized stormwater control technologies. Several of the decentralized controls are well suited for application in areas that are already planned to be vegetated. However, even when land is limited or seemingly non-existent, decentralized control options such as green roofs and infiltration planters, are still available. Areas with heavy industrial contamination of soils may not be suitable for infiltration strategies but are appropriate for decentralized controls that focus on water quality enhancements.

Land use and other considerations influence the practices selected and the operational characteristics. In some instances, such as when basement or foundation flooding is a concern or unfavorable or contaminated soils are present, infiltration of stormwater may not be desirable. In these instances other decentralized practices (e.g., stormwater storage) or a closed infiltration system may be a more appropriate solution. Existing utilities, structures, and environmental factors influence the use of infiltration in urban areas. Other locations, such as parks and riparian buffers may be ideal for stormwater infiltration. Because multiple hydrologic processes (e.g., retention, evapotranspiration, infiltration) are employed by decentralized controls, a number of treatment alternatives are available to be applied according to land use.

Often overlooked and underutilized, stormwater collection and reuse is an option that has significant potential. In addition to preventing runoff, cisterns and rain barrels provide a supply of water that can substitute for potable water in non-potable applications. For example, the 327,000-square-foot King Street Center in Seattle has more than 16,000 gallons of cistern storage capacity. Collected rainwater is filtered and pumped through a dedicated piping system for toilet flushing and irrigation needs. The cisterns are able to provide 60% (1.4 million gallons) of the toilet flushing water needed annually (NRDC, 2006). In addition, stormwater collection and reuse can take place on individual homeowner properties through the use of rain barrels. Several municipalities throughout the United States offer discounted rain barrels through grants. Water harvesting can take place on multiple scales; from single family homes, to multifamily properties, and larger commercial real estate. Stormwater collection and reuse can provide significant benefits to both the demand and discharge requirements of municipal water systems and integrate water management in a more holistic fashion. Some examples of integrated water management include

- ***Landscaped areas*** are often raised up and shed water. These areas can be reconfigured to function as part of a system of decentralized controls for stormwater control while still serving the original landscaped area functions.

- ***Vacant lots*** can be evaluated for use as part of the control strategy that a community adopts. The type of decentralized control used will be determined by the vacant lot's site-specific attributes that lend themselves to alternate solutions.

- ***Parks and playfields*** offer large land areas that can serve dual purposes. The recreational use is unimpeded by the introduction of a decentralized approach to stormwater control in this setting. Playing fields can be designed and built to act as infiltration areas for stormwater, forested areas in parks are natural infiltration zones, grassy swales can be used alongside of park roads and rain gardens can be installed as part of a habitat or environmental education initiative in a park. Other practices could be implemented as part of the control strategy for a given park or playfield.

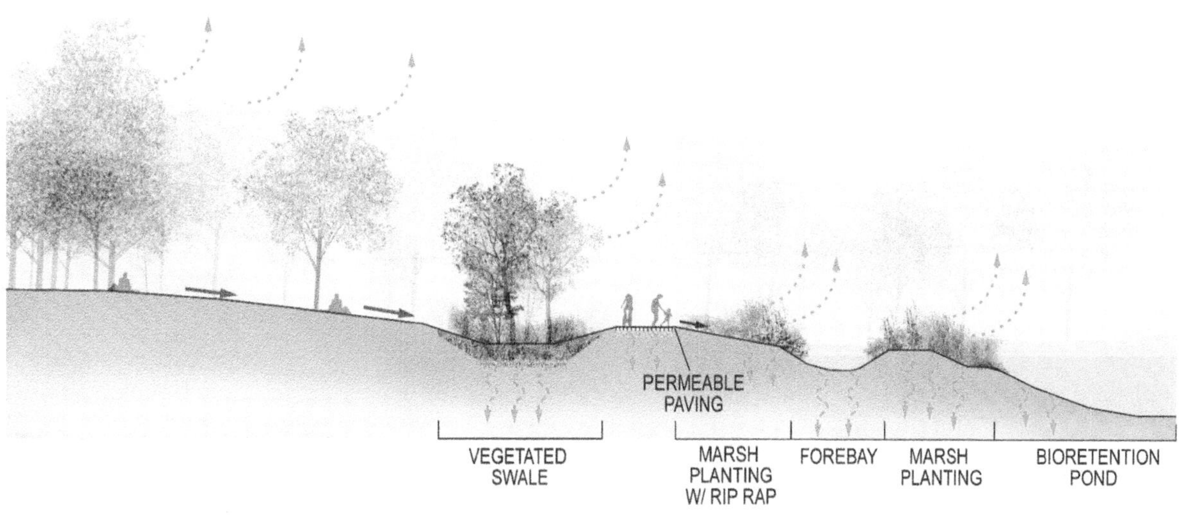

Figure 3-10. Typical Decentralized Controls for Parks and Open Space.

- ***Urban open space corridors*** can function as important linking elements in the decentralized control treatment train (Figure 3-10 and 3-11). These corridors are linear features that provide links between destinations for people, wildlife, and water. Several distributed retrofits are logical to evaluate for use in corridors. Grassy swales, rain gardens, and tree boxes are some of the retrofit tools that can be installed to mitigate stormwater runoff.

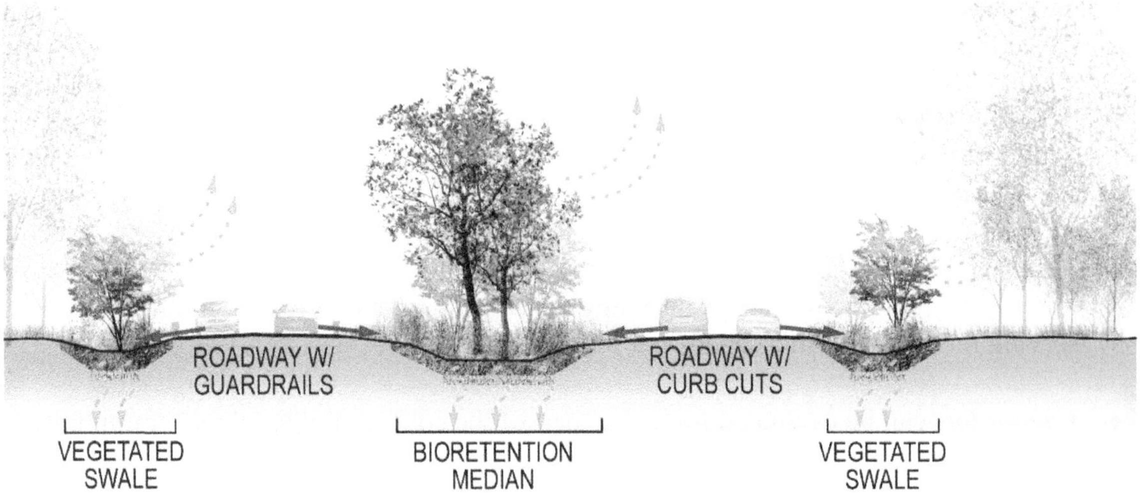

Figure 3-11. Typical Decentralized Controls for Open Space and Transportation.

- ***Commercial and governmental entrances, property line buffer zones, and campus applications*** provide retrofit opportunities (Figure 3-12). Entrances are developed to draw attention to a place; the use of a decentralized control that is attractive allows the entrance to serve the dual functions of marking the entrance and improving the stormwater quality and reducing stormwater volume. Rain gardens infiltrate and clean the water while creating an attractive low maintenance entrance. The style of a rain garden can vary from the very formal to an apparently wild aesthetic.

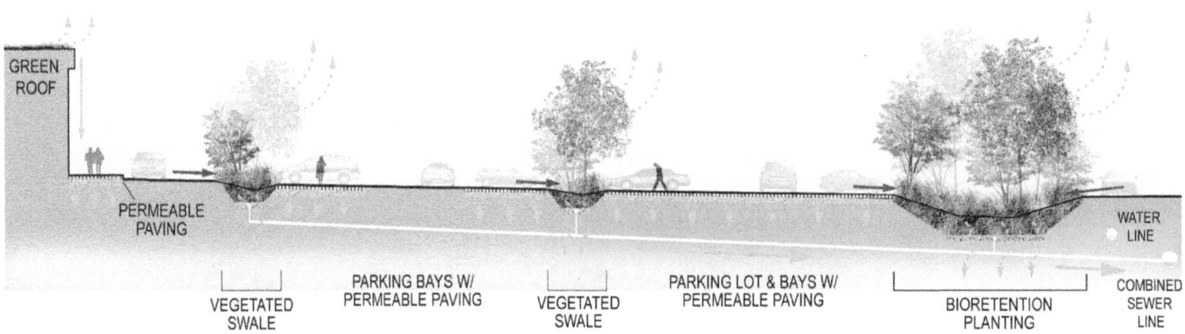

Figure 3-12. Typical Decentralized Controls for Government Buildings and Big Boxes.

♦ ***Urban residential areas*** provide many opportunities for decentralized control application (Figure 3-13). Almost the full range of control options are available for this land use type and are well suited as a stormwater control and aesthetic amenity. Typical building construction and often available green space, though limited, provide sufficient opportunity to integrate a treatment train approach that incorporates rainwater retention and reuse.

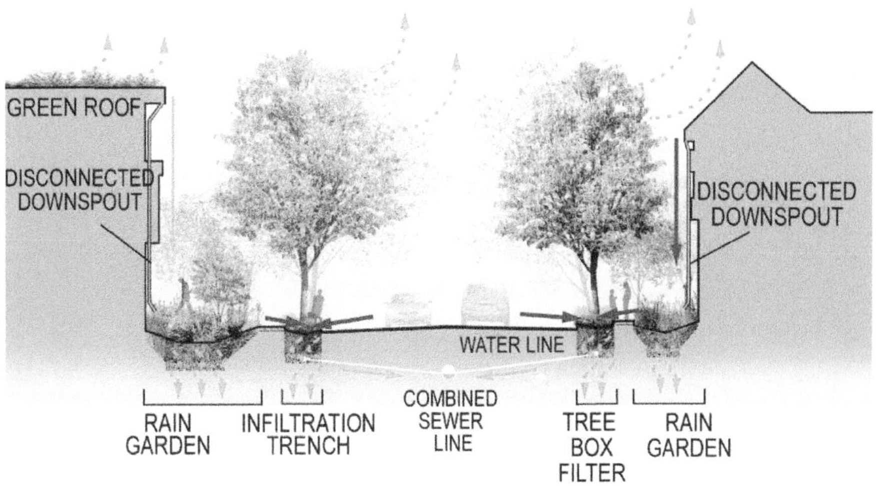

Figure 3-13. Typical Decentralized Controls for Urban Residential Areas.

♦ ***Suburban residential areas*** offer a range of possibilities for decentralized control application (Figure 3-14). A combination of preservation and minimization techniques can complement the introduction of decentralized controls. Available open space is conducive to downspout disconnection efforts, like Portland's. The use of decentralized controls also provides aesthetic and community benefits that have been well accepted by residents.

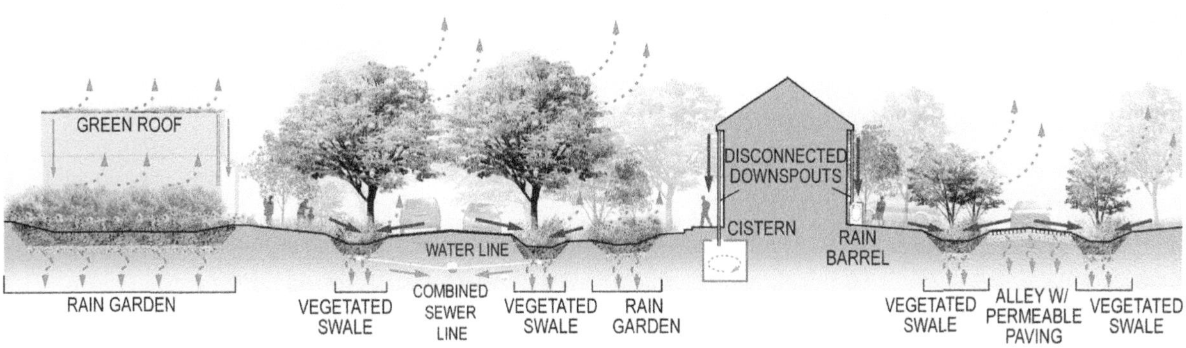

Figure 3-14. Typical Decentralized Controls for Suburban Residential Areas.

CHAPTER 4.0

FISCAL AND ECONOMIC BENEFITS OF DECENTRALIZED CONTROLS

4.1 Introduction

Managing stormwater and CSOs is one of the most pressing financing concerns in communities across the country. As the costs associated with maintaining traditional stormwater infrastructure increase, more communities are looking to alternative practices and decentralized controls to reduce the need for conventional treatment systems. Decentralized controls use the capacity of natural systems to reduce runoff volumes and increase or protect local water quality. Although the ecological and ecosystem benefits of these techniques have been fairly well established, the understanding of the associated economic value is still developing.

Local stormwater management has evolved from an urban flood control function to a water resource management function to an environmental protection and regulatory function. All three of these functions now co-exist and are the responsibility of local government officials. This has required a change in the way that stormwater management and mitigation efforts are planned, designed, constructed, managed, and perhaps most importantly, financed. Specifically, stormwater and CSO management has evolved from a basic capital construction and maintenance program supported primarily by local taxes to a program of integrated water resource management and environmental enhancement. As a result of these increasing complexities, community leaders and fiscal decision makers must develop multi-faceted, benefit-based financing systems (National Association of Flood and Stormwater Management Agencies, 2006).

In this chapter, the researchers evaluate the potential advantages of decentralized controls within the framework of the financing and investment process. By better understanding the economic and fiscal tradeoffs between conventional and decentralized BMPs, local and regional leaders can make more informed, effective allocation decisions. The analysis focuses on three key components of the financing process.

The first component is revenue. Managers responsible for watershed restoration need to leverage sufficient, sustainable revenues necessary to implement and maintain critical stormwater management programs. Money alone does not ensure success, but without sufficient revenue streams, implementation of many of the most critical BMPs will not occur.

The second key component is institutional capacity. Effective financing requires efficiency, and the complex nature of water infrastructure and water quality programs requires effective financing institutions. Many communities have explicitly made increasing institutional capacity a priority in their efforts to better allocate and manage fiscal resources.

Third, financing decision makers must develop investment strategies that maximize return on investment. Communities and public agencies charged with implementing and managing stormwater programs must justify their decisions, not only in terms of benefits to the natural environment, but also in terms of fiscal accountability and public support. They are being asked to demonstrate the economic and fiscal benefits of their investments. Even when it is impossible or impractical to measure benefits in dollars, agency staff can often provide evidence that their environmental investments are being managed to maximize environmental benefits per dollar spent (Ecosystem Valuation, 2007).

In addition, the various approaches to ecosystem valuation are described along with how they may be used to inform financing decisions related to paying for and implementing green infrastructure best management programs, practices, and techniques. The analysis focuses on the economic approach to valuation, its applicability to valuing ecosystem services, the limitations of these valuation tools, and the need for further study and investigation. Finally, case studies are provided that serve as examples of how decentralized controls are analyzed with newer methods of economic analysis.

4.2 The Financing Process

Faced with tightening budgets and multiple social, economic, and environmental needs, local leaders must make difficult decisions about how to allocate public investments. In making such decisions, communities must consider many objectives as part of investment decisions, including environmental quality, threats to ecosystem integrity, and the effects on people's quality of life (Ecosystem Valuation, 2007). To make these allocation and investment decisions effectively, sustainable and efficient financing and implementation strategies are necessary. While some communities have discovered that decentralized controls are often the most efficient options for mitigating the impacts stormwater and CSOs, the total value of these tools are usually not considered in the financing process. With a better understanding of the value of these approaches to stormwater management, as well as the tools necessary for gauging this value, decentralized controls could become as critical to the basic structure of local communities as traditional grey infrastructure.

Financing is an allocation process of acquiring, managing, and investing fiscal resources, and the purpose of the financing process is to maximize the value or return on investment. The process is universal – it applies to small and large projects, as well as public and private institutions. Financing is the not the reason for doing anything, including protecting the environment and water resources. Rather, finance provides a framework or a process for accomplishing goals in the most efficient way possible. As communities struggle to mitigate the effects of CSOs from combine sewer systems as well as controlling runoff from urban areas, they are faced with a significant financing dilemma.

Financing is predicated on two key activities: identifying and leveraging sustainable, dedicated revenue streams, and expending fiscal resources in a manner that maximizes program efficiency and the return on investment. By understanding the relationship between these financing components, as well as the tools, programs, and BMPs that affect them, community leaders and decision makers can develop and implement more effective and sustainable water quality and stormwater management programs.

4.2.1 Leveraging Revenue Streams

By any measure, the costs associated with mitigating the effects of both CSOs and stormwater runoff is significant. Nationally, the cost estimates for retrofitting combined sewer systems is estimated to be nearly $55 billion. Stormwater management will add another $9 billion to that. Regardless of the efficiency of the management programs developed to manage stormwater events, the costs are significant. For example,

- The mayor of Indianapolis, Indiana recently announced a $1.8 billion program to fix the city's CSO problem. To fund the program, which includes a 10-mile long tunnel, holding basins, and upgrades to the city's wastewater treatment plants, sewer rates will rise every year or two until 2025, to about $60 per month.
- To solve its CSO problem, Washington, DC, will construct four storage tunnels and related facilities at a cost of $1.9 billion. Though federal resources will assist in the financing, some ratepayers will see their bills increase by almost threefold.
- It is estimated that current state and local investment strategies to address all stormwater pollution issues throughout the Chesapeake Bay watershed will cost a total of $15 billion. About 60% of this cost, about $9 billion, is for retrofitting existing urban areas (Chesapeake Bay Finance Panel, 2004).

Financing programs and needs of this magnitude will require the participation of every citizen, taxpayer, ratepayer, and institution in the affected areas. And it will certainly require dedicated, sufficient revenue and fiscal resources.

Revenue is a term usually used in specific reference to the cash flow generated by user fees of various types, as well as other relatively consistent income streams. A revenue or financing source is the ultimate payer of a cost associated with any activity being paid for or financed. Public agencies and community leaders are always looking for new sources to fund restoration activities, and what makes financing urban stormwater management such a challenge is the complexity and difficulty of leveraging sufficient revenue sources. As a result of federal Clean Water Act requirements and the corresponding state implementation and enforcement programs, it is local municipalities and communities that are responsible for developing and implementing stormwater management programs. And, in fact, that is the way that it has always been. What has changed is the type of programs that taxpayers must now pay for.

One of the most significant implementation barriers facing communities is the lack of institutional capacity to finance stormwater and CSO management. Stormwater management has traditionally been financed through a variety of public agencies. This was often the most appropriate system, especially given the fact that most stormwater management plans focused on flood management and conveying water away from property as quickly as possible. In effect, pouring concrete was the most common stormwater management tool, either in the form of combined or separate storm sewer systems. However, as it became necessary to manage stormwater and CSOs based on water quality standards, many communities realized that a more efficient institutional arrangement was necessary. As a result, the development and implementation of enterprise-based programs, like those associated with wastewater and drinking water, are becoming much more common.

4.2.2 Financing Stormwater Management

The Water Quality Act of 1987 had a profound impact on how communities manage and pay for stormwater control programs. Before 1987, most communities viewed stormwater as primarily a flood management issue. The goal of most stormwater programs was to convey water away from property as quickly as possible. All that changed with the passage of the Water Quality Act, which required communities to consider the impact of stormwater on water quality. The law also provided a regulatory framework to ensure local and state compliance. As a result, the rules changed. The local requirements and responsibilities were significantly increased, while the responsibility for covering the costs associated with these increased responsibilities remained with local governments. The capacity of the financing sources, institutions, and investment tools all came into question, and communities increasingly turned to utilities or authorities to fund and finance stormwater programs.

The development of stormwater enterprise programs has been motivated by two primary financing needs. The first is related to revenue. A stormwater utility (or authority) is essentially a special assessment district set up to generate funding specifically for stormwater management. Users within the district pay a stormwater fee, and the revenue that is generated directly supports the maintenance and upgrade of existing storm drain systems; development of drainage plans, flood control measures, and water-quality programs; administrative costs; and sometimes construction of major capital improvements (Kaspersen, 2000). By collecting fees, communities ensure a stable source of revenue to finance the operations and maintenance of stormwater programs. And, unlike a stormwater program that draws exclusively on the general tax fund or uses property taxes for revenue, enterprise funds create a direct link between the polluter and the responsibility to pay.

The second reason for the development of stormwater enterprise programs is the need to gain administrative and fiscal efficiencies by coordinating and consolidating stormwater management programs. Often, communities implement various aspects of stormwater management programs within a variety of government agencies and programs, resulting in unnecessary redundancies and inefficient implementation of programs. Such inefficiencies increase costs and decrease the community's return on investment.

Though the development of stormwater enterprise programs has been an innovative and important step in the financing process, these programs are only part of the financing solution for mitigating stormwater runoff. The average monthly fee supporting stormwater enterprise programs across the country is between $3 and $4. In urban communities, fees at this level are only sufficient to finance operations and maintenance of existing infrastructure. By contrast Philadelphia and Portland, Oregon, cities that have adopted fees that are more reflective of actual stormwater program costs, have average monthly residential fees of $10 and nearly $17, respectively. In most communities, additional capital infrastructure requirements continue to be funded and financed with general tax revenue. Given that dedicated funding is limited, there is a pressing need to invest in BMPs that reduce program costs. Local stormwater managers are increasingly looking to decentralized controls as cost effective options for meeting stormwater management goals.

4.2.3 Financing CSO Management

Historically, the control and financing of CSO abatement has proven to be extremely complex. There are approximately 746 communities across the country served by combined systems, and the estimated cost of abatement is nearly of $55 billion (U.S EPA, 2008). Though U.S EPA has issued CSO control strategies that provide flexibility to those communities facing financial hardship, federal law will require all communities served by combined systems to mitigate the effects of these systems to stay in compliance with the federal Clean Water Act. Unlike stormwater, communities struggling with combined sewer systems have usually relied on existing wastewater financing institutions, such as wastewater systems, utilities, and authorities. Though the institutions are in place, the costs associated with controlling CSOs is significant, thereby placing a very real financing burden on municipal wastewater systems across the country.

One of the primary functions of a public water utility is to finance infrastructure engineering, construction, and maintenance. This starts with collecting sufficient revenues, and the source of this revenue is almost always the ratepayers within the system. In addition to funding traditional wastewater system construction, upgrades, and maintenance, communities with combined sewer systems are required to finance mitigation efforts to control overflow events. This has traditionally been accomplished by separating combined systems into two separate systems – by replacing the one combined pipe with two separated pipes – or by building holding and containment facilities that can control the volume of runoff flowing to wastewater treatment facilities. Both of these approaches are very expensive and have resulted in significant rate increases for utility customers. For example, Portland, Oregon, is in the 16th year of a 20-year, $1.4 billion, federally mandated project to reduce sewage overflows into the Willamette River from about 100 days a year to 4 days or less (Yardley, 2007). The work is financed through the city's wastewater treatment utilities. As a result, Portland has some of the highest water and sewer rates in the country; the average residential sewer bill has risen to about $45 a month from about $14 in the early 1990s, when the city began the mandated improvements (Yardley, 2007).

Mitigating CSOs presents a significant economic and fiscal burden to municipalities. As the costs and the responsibilities related to stormwater management increase, financing decision makers are looking for BMPs that can help reduce the costs associated with regulatory compliance. Decentralized controls have the potential to influence the financial costs of reducing CSOs. Although there has been a significant amount of interest in small-scale and decentralized environmental programming, many communities and financing institutions have not fully adapted decentralized BMPs.

Another aspect of evaluating the financing of CSO mitigation and LTCP appropriateness is the required affordability analysis. U.S. EPA developed the analysis procedures to assess the effect of CSO control costs on ratepayers and on the fiscal health of the community. Ratepayers typically experience a substantial increase in costs to cover combined sewer improvements. The requirement for the affordability analysis could provide further incentive for incorporating decentralized controls into CSO mitigation efforts because they may offer opportunity to limit the overall increase in utility rates.

4.2.4 How Decentralized Controls Affect Financing

As previously discussed, stormwater and CSO management has historically resulted in capital-intensive infrastructure projects. Municipalities have typically financed these projects by issuing debt. Paying the debt service on these projects usually is accomplished by increasing user rates or by allocating revenue from the general fund. However, decentralized controls have the potential to alter this conventional financing approach. Distributed control systems can be phased in over a number of years and can avoid the large upfront capital expenditures. This often means that debt issuance, and the interest payments over the life of the debt, can be avoided. This efficiency can help municipalities maximize infrastructure investments by paying less debt and either freeing more revenue for installation and implementation or limiting increases in user fees. Incremental financing also yields incremental environmental returns as the benefits of small-scale decentralized controls are realized shortly after expenditure. This is in contrast to large centralized projects that often take many years to provide benefit after investment.

Decentralized controls also alter the "who pays?" question. As more stormwater regulations emerge that require onsite stormwater management or the explicit use of decentralized controls, financing for a portion of stormwater management will transfer from the public to the private sector. Although public financing for stormwater management will always be required, decentralized controls can help to move closer to the "polluter pays" ideal.

Financing maintenance for decentralized controls will also differ from centralized systems. A distributed stormwater management system will require a coordinated maintenance program. And the number and type of decentralized controls may require additional maintenance and operational funding. Because many decentralized controls will be on private property, the private sector may incur some of the associated costs. The potential for additional maintenance costs and the potential for some of these costs to be offset by the private sector will have to be considered for financing purposes. How maintenance is handled is often a matter of municipal preference with some municipalities electing to shoulder much of the maintenance responsibility to not only ensure performance, but also the aesthetic appeal of the selected controls.

Various incentives are also available to be incorporated into the financing structure for decentralized controls. For instance, several municipalities allow for reductions in stormwater fees if decentralized controls are used by a rate payer. The fee reduction provides duel benefit by encouraging private entities to invest in onsite controls and obtain a utility fee reduction and by lessening the demand on the collection system. Similarly, tax incentives and grant funding have been used to offset a portion of the costs associated with installing and maintaining decentralized controls. Indirect financial incentives have also been used. Density bonuses or zoning variations have been provided for construction or redevelopment with decentralized controls. Other cities, such as Philadelphia and Chicago, provide expedited permit reviews for green projects that use decentralized controls, lessening the amount of time developers are leveraging debt (Table 4-1).

Table 4-1. Potential Economic Benefits and Challenges of Decentralized Controls.
Reprinted with permission from Wise (2007).

Benefits	Challenges
• Incremental implementation and funding can result in less debt service. • Decentralized controls are less capital intensive and may have overall lower costs. • Can extend existing capacity of current infrastructure. • Captures the asset values of clean water, soil capacity, and open space amenities: values ecosystem services. • Increases property values to the benefit of the private sector and public revenue collection.	• New financing mechanisms can create new administrative responsibilities within and across agencies. • Change in operational and maintenance needs and coordination with private sector.

4.3 Developing Effective Investment Strategies: The Role of Decentralized Controls

The development of efficient institutions capable of leveraging sufficient revenue sources is critical in the financing process. Ultimately, however, successfully mitigating the impacts of stormwater runoff and CSOs depends on the effectiveness of the BMPs, infrastructure, and programs in which communities invest. Because the costs associated with conventional stormwater management systems can be prohibitively high, the most effective investments are often associated with decentralized management approaches.

Studies documenting the effectiveness of decentralized controls have shown that these techniques can reduce infrastructure and construction costs in new development projects. U.S EPA's recent assessment (2007) found that total capital cost savings ranged from 15% to 80% when compared to conventional infrastructure. In addition, the flexible and distributed qualities of decentralized controls also allow them to be retrofitted into developed areas to provide site-specific stormwater and CSO control (NRDC, 2006). Nonetheless, wider adoption still faces obstacles. Among these is the economic investment that is required across the country for adequate stormwater and CSO control. Although decentralized controls are, in many cases, less costly than traditional methods of stormwater and sewer overflow control, the true costs and benefits associated with these BMPs have not been sufficiently studied. As a result, some municipalities continue to invest in conventional controls rather than trying these alternative approaches (NRDC, 2006).

The impetus for incorporating decentralized controls into stormwater management programs has often been based on the environmental benefits that these programs provide. In fact, these management tools are often the most effective way for communities to achieve water quality requirements and goals. It is the ability of natural systems to control water flow and filter pollutants that has lead to their adoption in many communities across the country. However, communities and public agencies charged with implementing and managing stormwater programs must justify their decisions, not only in terms of benefits to the natural environment, but also in terms of fiscal accountability. They are being asked to demonstrate the economic and fiscal benefits of their investments (Ecosystem Valuation, 2007). To truly understand the cost effectiveness and fiscal benefit of decentralized controls as stormwater and CSO management options, it is necessary to consider the total economic value associated with these tools. Though there is significant literature on the economic approaches necessary for determining the value of ecosystem or natural infrastructure services, it can be very difficult for local leaders to the acquire the necessary data and tools for making accurate valuation estimates.

4.3.1 The Total Economic Value of Decentralized Controls

Economic valuation can be defined as the attempt to assign quantitative and monetary values to goods and services provided by environmental resources or systems, whether or not market prices are available. When market prices are not available – for example when natural systems provide flood control services, or reduced erosion – the value is established by society's willingness to pay for the good or service, whether or not we actually make any payment (Lambert, 2003). Economic valuation is not an easy task, nor is it without conflict, which may explain why more communities have not implemented this type of analysis in determining the benefit of decentralized controls. These types of exercises often depend on what people perceive as the positive or negative effect that natural resources have on their wellbeing. In theory, the economic value of any good or service is measured in terms of what we are willing to pay for the commodity less what it costs to supply it (Lambert, 2003).

Despite these difficulties, however, there are good reasons for evaluating the benefit of natural resources and stormwater management approaches like decentralized controls. Local leaders must make difficult decisions about how to allocate public investments weighing environmental quality, threats to ecosystem integrity, and effects on people's quality of life (Ecosystem Valuation, 2007). As a result, it is not easy for decision makers to spend taxpayer's money on environmental activities, especially if there is no broad support from the public. Economic valuation is a way to estimate ecosystem benefits to people and allows financial experts to carry out a cost-benefit activity, which might support environmental investment. It is therefore an important tool for local decision makers to justify public spending on innovative management techniques (Lambert, 2003).

These types of valuation efforts are also important because people are not always aware of the values of natural resources in their communities. By providing objective evidence of the monetary and non-monetary benefits of natural resources, proponents of natural resource management and protection will gain public support. This is especially true in communities that are confronted with implementing and financing expensive stormwater management problems with limited fiscal resources. By better understanding the economic tradeoff between traditional and innovative BMPs, local and regional leaders can make more informed, effective allocation and investment decisions. For decentralized controls to be more widely adopted as acceptable stormwater management tools, the costs associated with these approaches, relative to conventional treatment, must be sufficiently calculated.

Economic valuation requires that ecosystems be described in terms of the goods and services that they provide to humans or other beneficiaries. Total economic valuation is based on the assumption that individuals and citizens can hold multiple values for ecosystems (National Research Council, 2004). In its simplest form, total economic valuation distinguishes between two components: use values and non-use values. The former refer to those values associated with current or future (potential) use of an environmental resource by an individual. In turn, use values are further distinguished by whether or not they have direct or non-direct value to citizens. Direct use values are benefits derived from commodities such as fish, agriculture, fuel wood, recreation, transport, wildlife harvesting, energy, etc. Although direct use values are limited in urban situations related to managing stormwater and CSOs, there are direct use values associated with green areas targeted for development. The most common values associated with decentralized controls and natural resources are indirect use values, which are indirect benefits derived from natural resource functions like nutrient retention, flood control, stormwater protection, and groundwater recharge. Much of the literature on the benefits of decentralized controls focuses on their associated indirect use values.

Non-use values associated with decentralized controls arise from the continued existence of the resource and are unrelated to use (National Research Council, 2004). Though these types of values are the most difficult to measure, they can be critically important to citizens, and can have a significant effect on the public's perception of urban environments. For example, Seattle piloted a Street Edge Alternatives (SEA Street) project to mitigate stormwater runoff from city streets by providing drainage that more closely mimics the natural landscape before development than traditional piped systems. In addition to the quantified environmental benefits, the project became so popular that residents from outside the project area have requested that similar projects be implemented in their communities. Although many citizens in Seattle care deeply about stormwater management in their city, it is reasonable to assume that these requests were being made as a result of perceived benefits that were in addition to the capacity of these tools to mitigate stormwater runoff. Quantifying the value of these benefits would be helpful as city leaders continue to develop their stormwater investment strategies.

Understanding the total economic value of decentralized controls is important to the investment process. Putting a monetary value on these natural systems, however, can be a challenge for local decision makers. The most direct way of assigning value is to rely on the market and the laws of supply and demand. This is appropriate for some ecosystem services, specifically those that have direct use value. However, decentralized controls do not usually provide direct benefit to citizens. As a result, the benefits that these tools do provide are often un-marketable. Even when we can place a market price on natural resource services, this price does not always reflect the true value that we place on them. In the Seattle SEA Street program example, it is reasonable to assume that many residents would be willing to pay more for the services provided by the decentralized practices than they were actually paying.

If monetary values of ecosystems services are not estimated, many of the major benefits will be excluded in the financing and resource allocation process. The likely outcome of such an omission would be too little protection of ecosystems and natural resources, and as a consequence the services that people directly and indirectly enjoy would be undersupplied. Valuation, therefore, can help to ensure that ecosystem services that are not traded in markets and do not have market prices receive explicit treatment in economic assessments financing processes (NRC, 2004). Given that the market does not provide the information that decision makers need, economists have developed additional tools, however imperfect, to assist in determining the total value of ecosystem services like those provided by decentralized controls.

4.3.2 Analyzing Cost Effectiveness

Decentralized control measures have the potential to influence the financial costs of mitigating CSOs and stormwater runoff. The costs associated with stormwater management programs are directly related to the effect of development practices on nature's ability to maintain a water balance (NRDC, 2006). Specifically, stormwater management programs are necessary to mimic nature's ability to control water flow and to filter pollutants from waterways. The more we adversely affect the natural water balance, the more we must implement increasingly expensive water quality programs. Distributed BMPs can help reduce the costs associated with stormwater management by restoring this balance, thereby decreasing the amount of pollution introduced into waterways and relieving the strain on stormwater and wastewater infrastructure (NRDC, 2006). Implementing decentralized control programs may be the most cost-effective and efficient way to achieve water quality goals and requirements.

Cost effectiveness analysis identifies the least expensive way of achieving a given environmental quality target, or the way of achieving the greatest improvement in some environmental target for a given expenditure of resources (Environmental Valuation, 2006). Perhaps the most direct benefit or value associated with decentralized control efforts is the cost savings and fiscal efficiencies that are generated. Communities across the country have realized cost savings and other fiscal benefits associated with reducing volume to CSO and stormwater infrastructure systems. Though there are many ways to measure cost effectiveness, this report focuses on four that are most appropriate when measuring the value of decentralized control programs: damage cost avoided, replacement cost, substitute cost, and lifecycle cost analysis.

4.3.2.1 Damage Cost Avoided, Replacement Cost, and Substitute Cost Methods

The damage cost avoided, replacement cost, and substitute cost methods are related methods that estimate values of ecosystem services based on either the costs of avoiding damages due to lost services, the cost of replacing ecosystem services, or the cost of providing substitute services. These methods do not provide strict measures of economic values, which are based on an individual's willingness to pay for a product or service. Instead, they assume that the costs of avoiding damages or replacing ecosystems or their services provide useful estimates of the value of these ecosystems or services. This is based on the assumption that, if individuals incur costs to avoid damages caused by lost ecosystem services, or to replace the services of ecosystems, then those services must be worth at least their replacement cost. Thus, the methods are most appropriately applied in cases where damage avoidance or replacement expenditures have actually been, or will actually be, made (Ecosystem Valuation, 2007).

Some examples of cases where these methods might be applied include:

- Researchers at the University of California at Davis have estimated that for every 1,000 deciduous trees in California's Central Valley, stormwater runoff is reduced nearly 1 million gallons – a value of almost $7,000. Preserving trees reduces polluted stormwater discharges and the need for engineered controls to replace those lost functions. When those trees are cut down and their functions are lost, those costs are passed on to municipal governments (NRDC, 2006).
- The District of Columbia Water and Sewer Authority (WASA, 2002) determined that the use of decentralized controls will help reduce the costs associated with managing

the city's CSO problem. Fiscal analysis comparing implementation scenarios with and without decentralized control measures indicated an overall cost savings with decentralized control use. In addition, WASA expects important secondary savings from these distributed retrofits. Stormwater diverted from the city's collection system reduces the flow to be processed at the city's wastewater treatment facility, which in turn could improve the capacity of the plant without the construction of additional facilities.

- Using its *CITYgreen* software program, the non-profit group American Forests has calculated the value of urban tree canopy for stormwater and CSO management in a number of cities and urban centers across the country. The calculations are based on the avoided construction costs associated with conventional stormwater management systems. The total volume of avoided water storage is multiplied by local stormwater and CSO retainment facility construction costs to determine the dollars saved by trees. For example, in the Delaware Valley region of Pennsylvania, it is estimated that tree canopy saves $2 per cubic foot of water for stormwater and $52 per cubic foot of water for CSO management (American Forests, 2003).

The damage cost avoided method uses either the value of property protected, or the cost of actions taken to avoid damages, as a measure of the benefits provided by an ecosystem. For example, if a wetland protects adjacent property from flooding, the flood protection benefits may be estimated by the damages avoided if the flooding does not occur or by the expenditures property owners make to protect their property from flooding (Ecosystem Valuation, 2007).

Because these methods use costs to estimate benefits, it is important to note that they do not provide a technically correct or total measure of economic value, which is properly measured by the maximum amount of money or other goods that a person is willing to give up to have a particular good, less the actual cost of the good. Instead, they assume that the costs of avoiding damages or replacing natural assets or their services provide useful estimates of the value of these assets or services. This method assumes that, if individuals incur costs to avoid damages caused by lost ecosystem services, or to replace the services of ecosystems, then those services must be worth at least what was paid to replace them. This assumption may or may not be true. However, in some cases it may be reasonable to make such assumptions, and measures of damage cost avoided or replacement cost are generally much easier to estimate than people's willingness to pay for certain ecosystem services. The methods are most appropriately applied in cases where damage avoidance or replacement expenditures have actually been, or will actually be, made (Ecosystem Valuation, 2007). This is the case with CSO and stormwater management and mitigation.

4.3.2.2 Lifecycle Cost Analysis

Most of the studies that support the use of decentralized approaches to stormwater management have focused on the construction or capital costs associated with various management alternatives. However, to accurately gauge cost effectiveness or efficiency of these systems, it is critical to consider costs over the entire life of the project. Lifecycle cost analysis (LCCA) is the tool used for calculating these costs.

LCCA is a method of project evaluation that considers all project costs arising from planning, designing, constructing, operating, and ultimately disposing of an asset (Powell et al., 2005). This type of analysis has been used extensively by federal, state, and local governments to evaluate the total cost of equipment purchases and maintenance. It is also a common tool for evaluating real estate investments and transportation projects. This approach to valuation is particularly suitable for the evaluation of stormwater and combined sewer system design alternatives that satisfy a required level of performance (including pollutant removal and stormwater retention rates) but that have different initial costs, operations and maintenance costs, and life spans. The lifecycle cost approach is critical in estimating decentralized control costs because often its operations and maintenance costs can be significantly lower than conventional approaches (Powell et al., 2005).

Though the potential "lifetime" value of decentralized controls could be significant relative to conventional approaches, the subjective nature of determining the time horizon for the project life cycle (such as 20 or 50 years) complicates LCCA. To fairly analyze the lifecycle cost of different alternatives, the estimator must use the same life cycle. In addition, the analysis should span the expected life of the major system components. Determining the reach of the costs considered also complicates LCCA. Some methods look at the broader societal costs, and others focus strictly on the direct project costs. Again, to fairly analyze the lifecycle cost of different alternatives, the estimator must use the same assumptions of cost reach for each (Powell, et al., 2005).

The issue of lifecycle cost could have significant effects on the adoption of decentralized controls. Though the infrastructure cost estimates related to CSO and stormwater management are significant, they do not approach the estimated costs associated with replacing existing aging water and wastewater infrastructure. In its 2005 "Report Card for America's Infrastructure," the American Society of Civil Engineers gave water and wastewater infrastructure across the country a D-minus and suggested it would take an investment of $390 billion to bring wastewater infrastructure alone up to par. U.S EPA has projected that unless cities invest more to repair and replace their water and sewer systems, nearly half of the water system pipes in the United States will be in poor, very poor or "life elapsed" status by 2020 (Yardley, 2007). Clearly, avoiding these types of expenses in the future should be a critical part of any stormwater management strategy, and calculating the long-term cost efficiencies associated with decentralized approaches deserves further investigation and study.

Cost analysis is the most direct way of determining the value of decentralized control programs, and determining cost efficiencies should be part of every community's investment strategy. Communities across the country have leveraged significant cost efficiencies by implementing distributed management programs. These case studies have shown that decentralized controls can help reduce infrastructure and development costs as well as minimize operations and maintenance costs (MacMullen, 2007). However, even in those situations where there are no marginal cost advantages associated with distributed approaches versus conventional systems, many communities still desire and demand the protection of natural resources and the restoration of local ecosystems through the use of decentralized controls. This implies that communities value these natural resources beyond their capacity to reduce costs.

From a community investment perspective, it is critical that the value of these benefits be incorporated into the decision making process. Because decentralized controls can offer more benefits than many conventional systems, cost-effectiveness analysis by itself fails to offer decision makers adequate information for evaluating the alternatives (MacMullen, 2007). Therefore, a more inclusive process of valuation must be used, one that accounts for all of the benefits and associated economic value that decentralized controls provide.

4.3.2.3 Benefit-Cost Analysis

The literature on stormwater management describes the cost advantage that decentralized controls can have over end-of-pipe or conventional stormwater or combined sewer systems. As discussed above, case studies that compare the cost-effectiveness of distributed controls and conventional systems report that decentralized approaches can help reduce infrastructure and development costs as well as minimize operations and maintenance costs. Some researchers, however, note that a lack of information on the additional economic benefits provided by decentralized approaches can ultimately impede adoption and implementation (MacMullen, 2007). Because decentralized controls can offer more benefits than conventional stormwater management systems, cost-effectiveness analysis fails to offer decision makers adequate information for evaluating the alternatives. Analysis based on benefits as well as costs (benefit-cost analysis) more accurately depicts the full range of environmental and economic benefits and costs of decentralized and conventional systems.

The BMPs associated with decentralized controls potentially provide myriad community benefits, many of which are not captured in cost efficiency calculations. For example, for efficient allocation decisions to be made, local leaders must understand the connections between issues such as decentralized controls and economic development; the effect of urban green spaces on human health; and the benefits of distributed practices such as green roofs on energy requirements. Strict cost efficiency calculations cannot provide sufficient information in these areas. Therefore, more robust valuation exercises are necessary.

The comparison of costs and benefits allows an explicit consideration of the trade-offs that are inevitably involved in most environmental policy decisions. It recognizes that achieving a particular objective or goal such as preservation of a particular ecosystem comes at a cost, because the resources that must be devoted to this preservation are not available for use in providing other goods and services. A typical benefit-cost analysis asks whether the benefits of preservation (or restoration) are "worth" the costs involved. In this sense, it ensures that the limited resources used to provide goods and services to society are used in the most efficient way – that is, to achieve the greatest net benefit (NRC, 2004).

4.3.2.4 Productivity Method

The productivity valuation method, also referred to as the net factor income or derived value method, is used to estimate the economic value of ecosystem products or services that contribute to the production of commercially marketed goods. It is applied in cases where the products or services of an ecosystem are used, along with other inputs, to produce a marketed good (Ecosystem Valuation, 2007).

For example, water quality affects the costs of treating municipal drinking water. Thus, the economic benefits of improved water quality can be measured by the decreased costs of providing clean drinking water (Ecosystem Valuation, 2007). The productivity method is most appropriate for this valuation calculation because this is a straightforward case where environmental quality directly affects the cost of producing a marketed good – in this case providing municipal drinking water. This example is one of the simplest cases, where cleaner water is a direct substitute for other production inputs, such as water purification chemicals and filtration, and it can be especially appropriate for communities that have CSOs that affect municipal water supplies. The 850 million gallons per year of stormwater and raw sewage that is released from CSOs contribute to elevated levels of pathogens, solid materials, debris, and toxic pollutants that can create significant public health and water quality concerns (University of North Carolina Environmental Finance Center, 2007). Therefore, reducing the impact of CSOs positively affects the quality of drinking water.

Stormwater runoff can also have a significant effect on the "production" of drinking water supplies, and in many communities, decentralized controls are critical components of water resource protection strategies. For example, the Schuylkill River is a primary drinking water source for more than 1.5 million people in the Philadelphia metropolitan area. As part of its source water assessment process, the Philadelphia Water Department identified stormwater, as well as pathogens from wastewater, as being the primary threats to drinking water supplies in the region. Though conventional treatment approaches will be part of the city's mitigation strategy, encouraging the adoption of decentralized controls in new development projects, and protecting open space have been identified as cost effective strategies for protecting the region's water resources (University of Maryland Environmental Finance Center, 2007).

4.3.2.5 The Hedonic Pricing Method

The hedonic pricing method is used to estimate economic values for ecosystem or environmental services that directly affect market prices. It is most commonly applied to variations in housing prices that reflect the value of local environmental attributes (Ecosystem Valuation, 2007). It can be used to estimate economic benefits or costs associated with environmental quality, including air pollution, water pollution, or noise; and, environmental amenities, such as aesthetic views or proximity to recreational sites. The basic premise of the hedonic pricing method is that the price of a marketed good is related to its characteristics, or the services it provides. For example, the price of a car reflects the characteristics of that car – transportation, comfort, style, luxury, fuel economy, etc. Therefore, we can value the individual characteristics of a car or other goods by looking at how the price people are willing to pay for it changes when the characteristics change. The hedonic pricing method is most often used to value environmental amenities that affect the price of residential properties (Ecosystem Valuation, 2007).

A recent study by researchers at the Wharton School of Business at the University of Pennsylvania provides an example of how the hedonic pricing method can be effectively used to value ecosystem services in urban communities. The study calculated the value of public investments in Philadelphia in "neighborhood greening," a general term to denote everything from adding parks to improving streetscapes to planting new trees in public spaces. The study showed that proximity to a greening event positively affects home values in the city. Proximity to a new tree planting is associated with overall increases in house prices of 9% (Wachter and Gillen, 2006).

Streetscapes are part of the "green infrastructure" of the urban environment. A streetscape project represents horticultural treatments to a sidewalk or roadway that improve the appearance of the area, making it a more attractive and pleasant place. Treatments can include tree plantings, container plantings, small pocket parks, parking lot screens, and median plantings. Streetscapes tend to focus on commercial corridors with high visibility and high levels of pedestrian and/or vehicular traffic. Their results indicate that streetscaping imparts a considerable increase in surrounding home values as well, approximately a 28% gain in value relative to similar homes in comparable areas without streetscape improvements (Wachter and Gillen, 2006).

Hedonic calculations can also be used to show the negative effect that degraded environments or natural resources can have on local communities. Studies have been done to show the impact of property values from nearby landfills, toxic waste sites, and water bodies impaired by CSOs. Although determining the total economic value of decentralized controls is cumbersome and difficult, the effects on local investment and resource allocation decisions could be significant. There is evidence to show that protecting and restoring the natural water balance can result in significant cost efficiencies associated with CSOs and stormwater management. In addition, decentralized controls can provide cost efficiencies related to other environmental management issues such as managing air quality. Again, however, the information needed to apply these values in the investment process is lacking (Table 4-2).

Table 4-2. Comparison of Costing Methods.

Method	Description	Positive Effects	Drawbacks
Damage cost avoided, replacement cost, and substitute cost methods	• Estimates values of ecosystem services based on the cost of avoiding damages due to lost services, replacing ecosystem services, or providing substitute services.	• Most useful and appropriate when actual damage or avoidance costs can be measured.	• Do not provide strict measures of economic values, but rather assume that avoided or replacement costs provide useful estimates of value.
Life-cycle cost analysis	• Considers all costs from planning and design through disposal.	• Good method for evaluating competing alternatives that have different initial costs, O&M costs, and life spans.	• Subjective nature of determining life span complicates analysis.
Benefit-cost analysis	• Considers benefits gained from investment as well as costs.	• More accurately reflects the full range of environmental and economic benefits and costs.	• Valuation of community and environmental benefits currently inadequate.
Productivity method	• Estimates the economic value of ecosystem services or products that contribute to commercially marketed goods.	• Can effectively estimate costs when treatment costs of environmental goods (e.g., water) are known.	• Only estimates the value of commercially marketable products.
Hedonic pricing method	• Estimates economic values for environmental services that directly affect market prices.	• Useful for estimating property value changes from use of decentralized controls.	• Limited to assessing effect on market prices.

4.4 Conclusion

Managing stormwater and CSOs is one of the most daunting tasks facing communities across the United States. Conventional methods of managing stormwater can be prohibitively expensive, and the use of decentralized controls practices for stormwater management present a number of potential fiscal advantages. Developing dedicated revenue streams for stormwater management and having the proper institutions in place to manage these resources are necessary and important first steps in the financing process; however, money alone will not ensure successful implementation of stormwater management plans. Having a clear understanding of the economic trade-offs that exist between conventional and decentralized approaches is critical if local decision makers are to allocate resources efficiently and invest specifically in the stormwater management practices and programs that enable communities to maximize the return on their investment.

4.5 Case Studies

Municipalities have begun using some of these alternative valuation methods to assess the economic implications of implementing decentralized controls. Seattle Public Utilities has formalized the process of using triple bottom line evaluations in its project development plans. Environmental and water quality concerns are of importance because of the public's connection to Puget Sound and the salmon population that it supports. SPU had historically considered environmental impacts in planning, but the triple bottom line evaluations have standardized and institutionalized the process and considerations. The triple bottom line evaluations are a hybridized environmental impact assessment that incorporates benefit-cost, present value, and budget analyses to quantify the social, environmental, and financial effects of a project. The goal of the evaluations is to reasonably assess alternatives in a more comprehensive manner than traditional economic assessments and to provide additional perspective and depth to the decision-making process (Seattle Public Utilities, 2005).

Toronto has used an academic study to assess the municipal level economic benefits of wide-scale green roof application. A study from Ryerson University used a geographic information systems database to model and aggregate environmental benefits and a method for monetizing these benefits to determine economic impacts. The study assumed that green roofs would be installed on all city flat roofs more than 3,750 square feet and cover at least 75% of the roof; these assumptions provided over 12,000 acres of green roofs representing 8% of the city's total land area. Five primary benefits were quantified in the study:

1. Reduced stormwater flows into the separate storm sewer system;
2. Reduced stormwater flows into the combined sewer system;
3. Improved air quality;
4. Mitigated urban heat island effects; and
5. Reduced energy consumption.

The economic assessment estimated that nearly $270 million in capital cost savings and more than $30 million in annual savings could be achieved with this approach. Of this amount, over $100 million was attributed to stormwater capital cost savings, $45 million to CSO capital cost savings, and $750,000 to CSO annual cost savings. With the average cost for a green roof in Canada approximately $6-7 per square foot (2005 U.S. dollars), each 3,750 square foot green roof would cost $22,000-$27,000. The total cost for all 12,000 acres would be $3 billion-$3.7 billion (Banting et al., 2005). While this investment is significant, it would be distributed across a number of public and private entities and years.

CHAPTER 5.0

MODELING DECENTRALIZED CONTROLS

5.1 Stormwater Modeling Approach

Whatever processes are used to integrate decentralized controls into capital projects, development projects, and/or planning, a stormwater model capable of simulating decentralized controls will be an essential part of any stormwater management program. Stormwater modeling is needed for defining the runoff problem, generating alternatives, evaluating alternatives, decision making, and long-term management.

This chapter discusses how models can be used to assess decentralized controls. The chapter provides both general and model-specific guidelines for the software modeling of seven unique decentralized BMPs. More specifically, the chapter focuses on four non-proprietary software models: U.S. EPA's Storm Water Management Model, version 5 (SWMM5), NRCS's Technical Release 55 (TR-55), U.S. EPA's Hydrologic Simulation Program - FORTRAN model (HSPF), and Western Washington's Hydrology Model, version 3 (WWHM3). There is a section for each model explaining its operations, how it addresses processes fundamental to decentralized controls, and a table outlining each model's approach to seven different controls. Details of each model's adeptness at modeling decentralized controls are also discussed. Although this chapter focuses mainly on modeling the water quantity aspects of the decentralized controls, each model's approach on key unit processes will also be addressed, albeit briefly and qualitatively.

Also, actual template files were created using SWMM5. The intended use of these templates is to provide a framework for the user in which to start the modeling process. A template exists for each of the seven controls listed in the BMP/Low Impact Development Model table. Each template is set up to run upon opening of the file under two different precipitation conditions. This way, the user can see the structure of a running SWMM model, and then make alterations to the template to adapt it to the project for which it is to be used. These alterations include, but are not limited to, changes in the physical characteristics of the subcatchment, physical characteristics of the BMP itself, and the storm events. These templates are explained in further detail in the template documentation in Appendix A.

Because decentralized controls are still an emerging practice and the water quality benefits of wide-scale application are not fully developed, models play a critical role in assessing the role of decentralized controls in stormwater management efforts. A Green Build-out Model was used to assess the impacts of city-wide green roof installation, urban reforestation, and tree box optimization on the separate and combined sewer systems of Washington, DC. Moderate and intensive build-out scenarios were modeled with a continuous simulation hydrologic and hydraulic model under annual rainfall conditions to determine the effects on runoff volume and discharge volume and frequency for the MS4 and combined systems. Some of the findings of the modeling effort are presented below (Casey Trees and LimnoTech, 2007).

- The intensive scenario reduced annual stormwater inflow into the sewer systems by more than 1.2 billion gallons. This resulted in a 10% reduction (more than 1 billion gallons) in discharges from the sewer systems and a 6.7% reduction in CSO frequency.
- The moderate scenario reduced inflow by more than 300 million gallons and reduced sewer discharges 3% and CSO frequency 1.5%
- Stormwater volume reductions of up to 7% were witnessed across the city.
- Reduction in discharges from the combined system was from 6% to 22%, depending upon the scenario.

The modeling effort and its results provide important information to decision makers weighing competing alternatives for stormwater management and allows for some quantification for the environmental benefits anticipated from a certain level of investment in decentralized controls. The Washington effort illustrates how modeling decentralized controls plays an important role in planning and coordinating how a municipality can employ decentralized controls most effectively.

5.2 Modeling Effort for Decentralized Controls

Modeling decentralized controls uses the same principles and processes as modeling conventional systems. An obvious realization is that instead of one BMP, several will be used, hinting that the scale of a decentralized modeling project can quickly become large in this regard. So ideally, the model must be capable of simulating large decentralized systems. Providing complex flow routing, explicitly addressing specific unit processes during each water balance calculation, and being able to physically represent the BMP structure with a high level of accuracy are also characteristics that an ideal model would exhibit for decentralized modeling.

In a centralized control system, the first destination of runoff is typically a conveyance device, virtually eliminating infiltration and evaporation. The opposite is true in a decentralized system. Decentralized controls emphasize treating stormwater close to the source, therefore promoting these two processes as soon as runoff begins. By promoting even more processes (e.g., interception, detention), decentralized controls can be implemented to meet multiple stormwater management objectives. This places special importance on these unit processes, as the ideal model would explicitly account for them. These BMPs also deliberately serve as stormwater filtration, and accounting for sedimentation (being a leading cause for BMP failure) would be a fitting characteristic as well.

Because decentralized controls treat stormwater at the source, the ability for a model to simulate runoff processes at a smaller, disaggregated scale now becomes important. In a residential development, this means simulating runoff at the parcel level. On the surface, this seems like a relatively easy ability for any model to provide. However, care must be taken as the equation systems that govern most conceptual models are based empirically on data sets from entire watersheds. The reality here is that the chance of a disaggregated simulation (many small land segments) producing the same runoff quantities as an aggregated simulation (one large land segment) is slim. The modeler must be aware and accepting of this margin of error, and ideally know enough about the model to minimize these errors as much as possible.

One of the important operations of decentralized controls is to retain the stormwater on site at the parcel level. A useful technique for this task is to use a downspout-disconnection method, or to sever the runoff link between roofs and the drainage system. This is done by routing the roof runoff over a pervious area of land to promote infiltration, interception, and evaporation. An effective way to simulate this in a model is to route the runoff from one land segment or roof segment onto another land segment.

Different sites will require different controls. It is highly likely that within one watershed, there exists one parcel that requires more runoff volume attenuation; whereas another may be producing less runoff, but contain more toxic contaminants. In addition, it is equally likely that physical site characteristics, such as soil compaction and slope, vary spatially. From this, the need arises to be able to assign different process parameters (e.g., infiltration and/or evaporation rates) to each BMP.

There are several models currently available that work exceptionally well at modeling common BMPs. And while it would be impossible to represent all of these models in a single document, the four represented herein have established a very significant history and application in the urban water resources industry. The model descriptions in Section 5.3 attempt to address the issues presented above.

5.3 Model Suitability

This section focuses on the implementation of four non-proprietary software models for use in modeling water quantities associated with decentralized stormwater controls. The models discussed in this section are SWMM5, TR-55, HSPF, and WWHM3. For reference, the graphical user interfaces of each model are pictured in Appendix B.

These four models were chosen for this evaluation because of several factors. SWMM5 is perhaps the predominant urban drainage model in use in the United States and is currently enjoying continuous support and updating from the U.S. EPA. SWMM in general dates from 1971, has a large user base and has often been used for wet-weather control simulation. NRCS hydrologic techniques are arguably the predominant routines used in hydrologic engineering practice in the United States. For example, NRCS loss routines involving curve numbers are included as an option in SWMM5. TR-55 is the NRCS computer model employing these routines, and some LID modeling in the Washington, DC, area is based on TR-55. Another principal American hydrologic model is HSPF, supported both by the U.S. EPA and by the U.S. Geological Survey. It is the required model for stormwater impact evaluation in the Puget Sound region and includes a recent toolkit specifically for simulation of BMPs. WWHM3 is based on HSPF for use in Western Washington and was found to be convenient enough to use to warrant inclusion as an option in this study.

Presented in this section is an overview of each model, how the model simulates the vertical water balance (infiltration and evaporation), and methods in which the model may be used in the simulation of decentralized controls. Accompanying each model description is a table containing the model's approach to seven representative structural BMPs, broken down by how the model simulates inflow, outflow, and losses.

5.3.1 U.S. EPA Storm Water Management Model Version 5 (SWMM5)

5.3.1.1 Model Overview

SWMM was first created in 1971 as a dynamic rainfall-runoff simulation program whose intended use was primarily for simulation of CSO controls. Since then, it has undergone several updates from both public and private agencies to include many more components of the total urban hydrologic cycle. SWMM is able to provide runoff quality and quantity as an aid in the design of a variety of conveyance systems (pipes, channels, pumps, and regulators) and storage devices. SWMM accounts for the major processes of the hydrologic cycle, including precipitation, evaporation, snow accumulation and melt, infiltration, groundwater interactions, and interception. The latest version of SWMM5 is available for download from the U.S. EPA website (http://www.epa.gov/ednnrmrl/models/swmm/index.htm).

Simulation of these parameters is performed within SWMM by the interaction of four "environmental compartments" (Rossman, 2005). SWMM requires the user to fabricate a conceptual, networked system of land segments, nodes and links, and specify precipitation and groundwater characteristics. Within these compartments, SWMM is able to compute stormwater runoff quantities and many water quality processes, such as buildup, wash-off, and first-order decay. The four environmental compartments are:

- **Atmosphere:** Using Rain Gage objects, the atmosphere compartment interacts directly with the land surface compartment, depositing precipitation and pollutants.
- **Land Surface:** Expressed as Subcatchment objects, the land surface compartment receives precipitation from the atmosphere compartment. This compartment then dictates how precipitation is divided between infiltration (to the groundwater compartment) and runoff (to the transport compartment).
- **Groundwater:** Using Aquifer objects, the groundwater compartment simulates subsurface interactions by transferring infiltration to the transport compartment.
- **Transport:** The transport compartment is composed of conveyance and storage objects, which obtain inflow from the groundwater and land surface compartments.

SWMM uses a water balance (illustrated in Figure 5-1) for subcatchments to calculate the runoff and losses that occur when modeling a land segment. SWMM treats each land segment as a non-linear, vertically-sided reservoir, subject to depression storage (D_p). The depth of water over the land segment is calculated continuously with time by solving the water balance (Huber and Dickinson, 1968):

$$\frac{dD}{dt} = I - E - F - Q \tag{5.1}$$

where D is the depth of water over the land segment (ft), I is the precipitation rate (ft/sec), E is the evaporation rate (ft/sec), F is the infiltration rate (ft/sec), and Q is the rate of runoff (cfs/ft^2 or ft/sec).

Runoff from the land segment occurs only when the depression storage is exceeded. Once this condition exists, SWMM calculates Q from the equation (Huber and Dickinson, 1968):

$$Q = \frac{1.49}{n}\frac{W}{A}(D - D_p)S^{0.5} \quad (5.2)$$

where n is the Manning's coefficient, W is the land segment width (ft), A is the area of the land segment (ft^2), D_p is the depth of depression storage (ft), and S is the slope of the land segment.

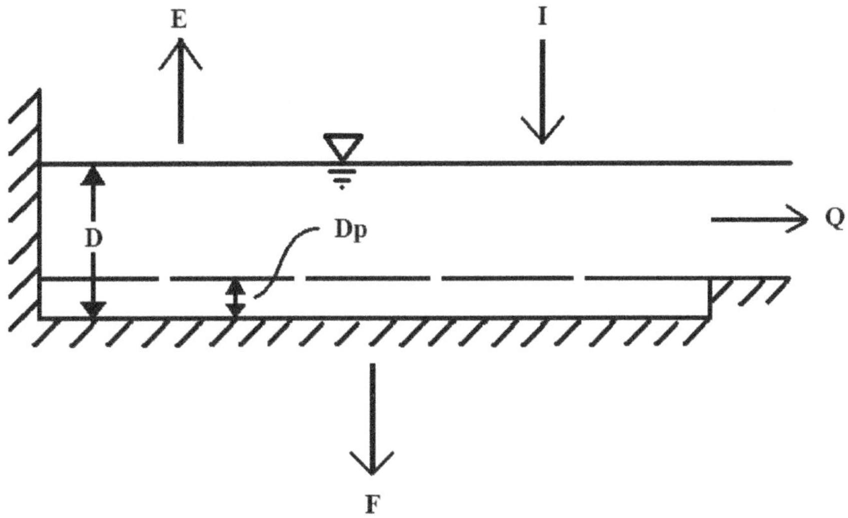

Figure 5-1. SWMM's Water Balance for a Land Segment (Rossman, 2005).

Runoff from one land segment may be routed onto either another land segment, or to a conveyance system of links and nodes used to simulate conduits and channels. Flow through these link and node systems within SWMM is governed by the conservation of mass and momentum equations for gradually varied, unsteady flow (i.e., the Saint Venant flow equations) or by the kinematic wave method. SWMM makes available three routing techniques to the user:

- **Steady Flow Routing:** This method is the simplest form of routing, as it translates the link's inflow hydrograph to the outflow hydrograph with no delay or change in shape. This uses Manning's equation to relate flow rate to water depth.
- **Kinematic Wave Routing:** This method solves both the continuity equation and a simplified momentum equation for each link. The maximum flow that can be conveyed through a link is the Manning equation value at maximum depth. Flow in excess of this value is either eliminated entirely from the system, or is "ponded" at the top of the inlet node and re-introduced as soon as capacity is available. This method allows for delays and changes from the inflow hydrograph to the outflow hydrograph.
- **Dynamic Wave Routing:** This method is the most advanced and solves the complete Saint Venant equations of continuity and momentum. This method can account for link storage, backwater, flow reversal, and pressurized flows and surcharging during events in which the maximum Manning equation flow is exceeded.

Flow routing through storage is performed by a Runge-Kutta numerical solution of the continuity equation and applicable outflow equations (e.g., rating curves) for the two unknowns, storage and outflow. The applicable outflow equations are dependant upon the presence of a weir, orifice, pump curve, table, or any combination thereof.

5.3.1.2 Infiltration, Evaporation, and Sedimentation

Evaporation is simulated for all nodes and subcatchments as either a constant value, as a time series, or as monthly average values. These values are applied to the water balance above (for subcatchments) or, if applied to a node, are simply subtracted from the quantity of water flowing through the node. SWMM simulates infiltration occurring on subcatchments in one of three ways (Bedient et al., 2008; Huber, 2008):

- **Horton Infiltration:** This method was created based on empirical observations showing that the rate of infiltration decreases exponentially from a maximum value to a minimum value over the course of a rainfall event. Necessary parameters for this method include the maximum and minimum infiltration rates, a decay constant that describes how fast the infiltration decreases, and the time it takes for the fully saturated soil to dry. Although easy to comprehend, Horton parameters are not always readily available.
- **Green-Ampt Infiltration:** This method assumes the presence of a sharp wetting front in the soil column that separates soil with saturated water content (above) from soil with only some moisture (below). Necessary parameters for this method include initial moisture deficit, hydraulic conductivity, and the wetting front's suction head. While more complicated than Horton, Green-Ampt parameters are readily available (Rawls et al., 1983).
- **Curve Number Infiltration:** This method classifies soils into groups that relate infiltration capacity to an NRCS Curve Number. As a rain event progresses, infiltration capacity is depleted as a function of cumulative rainfall and remaining capacity. Necessary parameters for this method include the curve number, hydraulic conductivity, and the time it takes for the fully saturated soil to dry.

Infiltration occurring in storage units is simulated using a time series of negative values. Alternatively, one may also use an outflow rating curve, which relates the amount of outflow to the head in the storage unit. If properly implemented, outflow rating curves can account for changes in the amount being infiltrated based on the water level in the storage unit (head).

To simulate infiltration occurring in conduits, a time series of negative values must be applied to the node before or after the conduit itself.

SWMM simulates sedimentation buildup and removal on the land surface, as well as treatment within storage units. Storage unit treatment is handled via a user-defined mathematical expression to calculate post-treatment concentrations or fraction removals. Buildup and removal characteristics on subcatchments are dependant upon the land uses assigned to those areas. These techniques apply not only to sediment, but other pollutants as well, but are not necessary for most decentralized control simulation efforts that focus on the water balance.

5.3.1.3 Modeling Decentralized Controls

Modeling BMPs in SWMM requires the use of three objects. Storage unit nodes, subcatchments, and conduits are used to model the seven structural BMPs listed in Table 5-1.

Storage unit nodes can be used for any BMP that requires water storage. Geometry is specified using a depth-area relationship, and the user is able to specify a multiple-orifice and/or weir outlet structure. Utilizing a subcatchment that is 100% pervious to receive runoff from adjacent areas will allow modeling of infiltration trenches, porous pavements, grass swales, dry wells, and any other BMP that is infiltration intensive. Conduits can be used to model BMPs in which water velocities are deemed an important design factor. Certain swales or wetlands, for instance, may be an appropriate application for using a conduit or system of conduits.

SWMM's strength in decentralized BMP modeling lies in its versatility. There is no limit to the quantity of BMPs or subcatchments for use in a disaggregated simulation. Its flow routing capabilities are equally adept, allowing the runoff from one subcatchment onto another. Also, SWMM allows for a high rate of precision by incorporating numerous possibilities for simulating infiltration and evaporation. In decentralized control modeling, these parameters will vary based on BMP type and spatial conditions, and SWMM is able to account for this.

Template files were created in SWMM5 for the BMPs in Table 5-1. For more information regarding modeling BMPs in SWMM, refer to the template documentation presented in Appendix B.

Limitations:

- SWMM's infiltration methods are only applied to subcatchments, and not conduits or storage units.

Table 5-1. SWMM5 BMPs.

<table>
<tr><th>BMPs</th><th colspan="2">SWMM 5</th></tr>
<tr><td>Ponds</td><td colspan="2">Modeled as a "Storage Unit" within SWMM

INFLOW:
runoff from upstream source (input of precipitation time series and/or continuous flows)

OUTFLOW:
- solved using Runge Kutta method
- lumped continuity and inflow/outflow relationships for flow routing
- rating curves (flow vs. depth)
- multiple outlets allowed, including weirs (low flow, overflow, etc.)

LOSSES:
ET and infiltration simulated by the input of a time series of negative flows</td></tr>
<tr><td rowspan="2">Wetlands and bioretention facilities</td><td colspan="2">Modeled as a "Storage Unit"
OR
Modeled as a "Conduit" within SWMM</td></tr>
<tr><td>Storage Unit
INFLOW:
same as ponds

OUTFLOW:
same as ponds

LOSSES:
same as ponds</td><td>Conduit
INFLOW:
same as ponds

OUTFLOW:
Solved using Manning's equation

LOSSES:
same as ponds</td></tr>
</table>

Table 5-1. SWMM5 BMPs (continued).

<table>
<tr><td rowspan="2">Infiltration trenches</td><td colspan="2">Modeled as a "Storage Unit"
OR Modeled as a 100% pervious "Subcatchment" downstream of a runoff source subcatchment</td></tr>
<tr><td>Storage Unit
INFLOW:
same as ponds

OUTFLOW:
same as ponds

LOSSES:
same as ponds</td><td>Subcatchment
INFLOW:
same as ponds

OUTFLOW:
- solved via a water balance
- treated as a rectangular open channel once depression storage is exceeded
- low slope and high Manning's n value needed to eliminate horizontal outflow from trench unless some outflow is desired

LOSSES:
ET simulated with negative time series and infiltration simulated using the Horton, Green-Ampt or Curve Number method</td></tr>
<tr><td>Grass swales and filter strips</td><td colspan="2">Modeled as a 100% pervious "Subcatchment" routed downstream of a runoff source

INFLOW:
same as ponds

OUTFLOW:
- solved via a water balance
- treated as a rectangular open channel once depression storage is exceeded
- low slope and high Manning's n value needed to eliminate horizontal outflow from swale/strip unless some outflow is desired

LOSSES:
ET simulated with negative time series and infiltration simulated using the Horton, Green-Ampt or Curve Number method</td></tr>
<tr><td>Dry wells</td><td colspan="2">Modeled as a "Storage Unit"

INFLOW:
same as ponds

OUTFLOW:
same as ponds

LOSSES:
- ET simulated with negative time series
- Infiltration simulated with outflow rating curve (to reflect infiltration changes with water depth, as dry wells are typically deep with minimal surface area)</td></tr>
</table>

Table 5-1. SWMM5 BMPs (continued).

Cisterns	Modeled as a "Storage Unit" INFLOW: same as ponds OUTFLOW: same as ponds LOSSES: ET and water use schedule simulated with negative time series
Porous Pavement	Modeled as a "Subcatchment" INFLOW: same as ponds OUTFLOW: - solved via a water balance - treated as a rectangular open channel once depression storage is exceeded LOSSES: ET simulated with negative time series and infiltration simulated using the Horton, Green-Ampt or Curve Number method

5.3.2 NRCS Technical Release 55 (TR-55)

5.3.2.1 Model Overview

TR-55 is a single event, rainfall-runoff model released originally in 1975 by the Natural Resources Conservation Service (formerly called the Soil Conservation Service, or SCS). TR-55 is applicable to small urban and agricultural watersheds, and is used primarily for generating peak stormwater runoff volumes, storage volumes, and hydrographs (NRCS, 1986). The latest version of TR-55 is available for download from the NRCS website (http://www.wsi.nrcs.usda.gov/products/W2Q/H&H/Tools_Models/WinTR55.html).

TR-55 uses a system of SCS-generated equations that provide a relationship, based on a curve number (CN), between the amount of precipitation and direct runoff. The SCS runoff equation is:

$$Q = \frac{(P - I_a)^2}{(P - I_a) + S} \tag{5.3}$$

where Q is the direct runoff depth (inches), P is precipitation depth (inches), I_a is the initial abstraction depth (inches), and S is the potential abstraction or storage depth (inches).

The initial abstraction term (I_a) accounts for all losses that occur before actual runoff begins. It includes water that is retained in depression storage and interception, and water lost in the evaporative and infiltrative processes. I_a is related empirically to the potential abstraction (S) by the relationship

$$I_a = 0.2S \qquad (5.4)$$

The runoff curve number, which is related to land uses and soil type, is related to potential abstraction (S) by the relationship

$$S = \frac{1000}{CN} - 10 \qquad (5.5)$$

The SCS methods incorporate synthetic rainfall distributions with which to influence the runoff hydrograph from a land segment. These distributions relate any time during the duration of a storm to its fraction of rainfall. Because rainfall intensities vary geographically, four synthetic rainfall distributions were created for different regions of the United States (I, IA, II, and III) and are shown in Figure 5-2. Type I and IA represent the Pacific coast climates with wet winters and dry summers. Type III represents the regions along the Gulf of Mexico and Atlantic coast where tropical storms bring considerable 24-hour rainfall amounts. Type II represents the rest of the country.

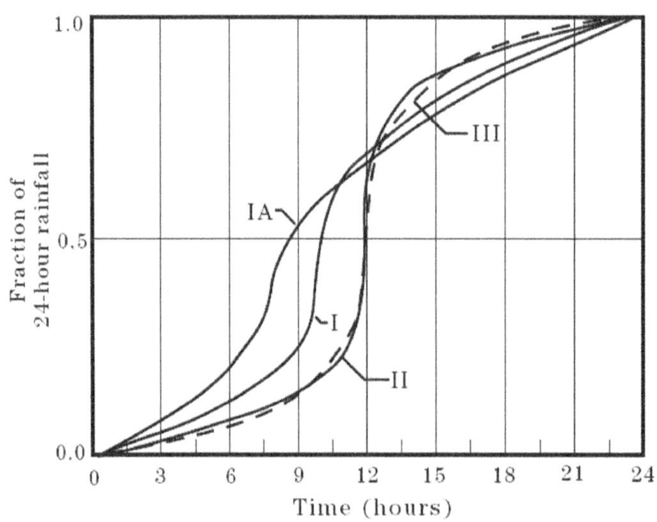

Figure 5-2. SCS-Generated Synthetic Rainfall Distributions (NRCS, 1986).

The shape of the hydrograph and its peak are also affected by the time of concentration (T_c), which is the time required for water from the most hydraulically distant point of the watershed to contribute to flow at the outlet. The T_c calculation requires that a water wave travel time (T_t) be calculated for each individual flow segment within the watershed. Travel time, the ratio of flow length to wave velocity, is given by the relationship

$$T_t = \frac{L}{3600V} \quad (5.6)$$

where L is the flow length and V is the average wave velocity, e.g., kinematic wave or dynamic wave, based on surface properties and conveyance type.

These travel times are summed to calculate the total time of concentration:

$$T_c = T_{t1} + T_{t2} + \ldots T_{tm} \quad (5.7)$$

5.3.2.2 Infiltration, Evaporation, and Sedimentation

Both infiltration and evaporation are simulated in the initial abstraction term (Ia), and are therefore depend heavily on the curve number. The NRCS has developed a table of curve numbers related to land use and the soil's hydrologic soil group, shown in Table 5-2. Soil group A is sandy and well drained, group B is sandy loam, group C is clay loam or shallow sandy loam, and group D is a poorly drained plastic clay (Bedient et al., 2008). Sedimentation, or any pollutant loading, is not simulated in the TR-55 model.

Table 5-2. Curve Numbers (NRCS, 1986).

Cover description		Curve numbers for hydrologic soil group			
Cover type and hydrologic condition	Average percent impervious area 2/	A	B	C	D
Fully developed urban areas (vegetation established)					
Open space (lawns, parks, golf courses, cemeteries, etc.) 3/:					
Poor condition (grass cover < 50%)		68	79	86	89
Fair condition (grass cover 50% to 75%)		49	69	79	84
Good condition (grass cover > 75%)		39	61	74	80
Impervious areas:					
Paved parking lots, roofs, driveways, etc. (excluding right-of-way)		98	98	98	98
Streets and roads:					
Paved; curbs and storm sewers (excluding right-of-way)		98	98	98	98
Paved; open ditches (including right-of-way)		83	89	92	93
Gravel (including right-of-way)		76	85	89	91
Dirt (including right-of-way)		72	82	87	89
Western desert urban areas:					
Natural desert landscaping (pervious areas only) 4/		63	77	85	88
Artificial desert landscaping (impervious weed barrier, desert shrub with 1- to 2-inch sand or gravel mulch and basin borders)		96	96	96	96
Urban districts:					
Commercial and business	85	89	92	94	95
Industrial	72	81	88	91	93
Residential districts by average lot size:					
1/8 acre or less (town houses)	65	77	85	90	92
1/4 acre	38	61	75	83	87
1/3 acre	30	57	72	81	86
1/2 acre	25	54	70	80	85
1 acre	20	51	68	79	84
2 acres	12	46	65	77	82

5.3.2.3 Modeling Decentralized Controls

A watershed is composed of sub-areas and reaches in TR-55. Reaches are designated as either channel or structure reaches. Modeling any facilities that require storage is performed using these structure reaches. The user specifies a depth-area relationship to shape the structure reach and has the option of selecting either a pipe or weir spillway type. Multiple outlets are not supported. TR-55 is able to report a volume and stage within the structure reach for each storm event during the peak flow period only. This is the extent of storage modeling in TR-55 (NRCS, 2002).

Modeling infiltration-dependent BMPs (porous pavements or filter strips, for example) is more of a gray area. Intuitively, modeling a porous pavement as a small sub-area with low CN would seem adequate, but the user should take care here. Routing constraints within the program prohibit the runoff from one sub-area from moving onto another. Therefore, the only water influx to a sub-area is from the user-defined precipitation events. A "creative" solution to this is to run multiple simulations. One simulation would be used to determine the volume of water that would be flowing to the porous pavement as runoff. The precipitation event in the separate porous pavement simulation could subsequently be altered to reflect the runoff quantity. However, the validity of manipulating this model in this way has not been evaluated, but the option remains if the modeler was determined to use TR-55. Similar work-arounds would be required for modeling the other BMPs in Table 5-3 that function with both storage and infiltration (drywells and infiltration trenches), because TR-55 does not apply losses to structure reaches.

Limitations:

- Curve numbers describe average conditions and do not allow changes with time.
- Routing constraints prohibit runoff from one sub-area onto another.
- Sedimentation is not simulated.
- Single storm event simulation only, continuous simulation not supported.
- Disaggregated simulations are limited to only 10 separate sub-areas.
- Vertical water balance not applied to structure reaches.

Table 5-3. TR-55 BMPs.

BMPs	TR-55
Ponds	Modeled as a "Structure" reach within TR-55 INFLOW: runoff from upstream source (SCS method of entering storm event magnitudes) OUTFLOW: Solved using Bernoulli's equation (if using a pipe as the spillway type) Solved using weir equations (if using a weir as the spillway type) LOSSES: losses are simulated in the initial abstraction (Ia) and Curve Number (CN) terms of the SCS equations for sub-areas. Losses cannot be applied to the pond itself.

Table 5-3. TR-55 BMPs (continued).

<table>
<tr>
<td>Wetlands and bioretention facilities</td>
<td>Modeled as a "Structure" reach

INFLOW:
same as ponds

OUTFLOW:
same as ponds

LOSSES:
same as ponds</td>
</tr>
<tr>
<td>Infiltration trenches</td>
<td>Modeled as a "Structure" reach
OR
Modeled as a "Sub-area" within TR-55

Structure reach
INFLOW:
same as ponds

OUTFLOW:
same as ponds

LOSSES:
same as ponds

Sub-area
INFLOW:
storm event magnitudes (2-yr, 5-yr, 10-yr, etc.)

OUTFLOW:
solved using SCS method

LOSSES:
Losses are simulated in the initial abstraction (Ia) and Curve Number (CN) terms of the SCS equations.</td>
</tr>
<tr>
<td>Grass swales and filter strips</td>
<td>Modeled as a "Structure" reach
OR
Modeled as a "Sub-area" within TR-55

Structure reach
INFLOW:
same as ponds

OUTFLOW:
same as ponds

LOSSES:
same as ponds

Sub-area
INFLOW:
storm event magnitudes (2-yr, 5-yr, 10-yr, etc.)

OUTFLOW:
solved using SCS method

LOSSES:
Losses are simulated in the initial abstraction (Ia) and Curve Number (CN) terms of the SCS equations.</td>
</tr>
</table>

Table 5-3. TR-55 BMPs (continued).

Dry wells	Modeled as a "Structure" reach OR Modeled as a "Sub-area" within TR-55 **Structure reach** INFLOW: same as ponds OUTFLOW: same as ponds LOSSES: same as ponds	 **Sub-area** INFLOW: storm event magnitudes (2-yr, 5-yr, 10-yr, etc.) OUTFLOW: solved using SCS method LOSSES: Losses are simulated in the initial abstraction (Ia) and Curve Number (CN) terms of the SCS equations.
Cisterns	Modeled as a "Structure" reach OR Modeled as a "Sub-area" within TR-55 **Structure reach** INFLOW: same as ponds OUTFLOW: same as ponds LOSSES: same as ponds	 **Sub-area** INFLOW: storm event magnitudes (2-yr, 5-yr, 10-yr, etc.) OUTFLOW: solved using SCS method LOSSES: Losses are simulated in the initial abstraction (Ia) and Curve Number (CN) terms of the SCS equations.
Porous Pavement	Modeled as a "Structure" reach OR Modeled as a "Sub-area" within TR-55 **Structure reach** INFLOW: same as ponds OUTFLOW: same as ponds LOSSES: same as ponds	 **Sub-area** INFLOW: storm event magnitudes (2-yr, 5-yr, 10-yr, etc.) OUTFLOW: solved using SCS method LOSSES: Losses are simulated in the initial abstraction (Ia) and Curve Number (CN) terms of the SCS equations.

5.3.3 U.S. EPA Hydrological Simulation Program – FORTRAN (HSPF)

5.3.3.1 Model Overview

The HSPF model's origins date back to the early 1960s and the Stanford Watershed Model. Beginning in the 1970s, the model went through several iterations and updates to include water quality processes and overall code and algorithm enhancements (Bicknell et al., 2001). The latest release of the interactive Windows version of HSPF, WinHSPF, is version 12, available as part of the U.S. EPA BASINS model (http://www.epa.gov/waterscience/basins/). Earlier releases of HSPF by itself are available from the U.S. EPA website www.epa.gov/ceampubl/swater/index.htm and the USGS website http://water.usgs.gov/software/lists/surface_water/. Incorporation of HSPF into the U.S. EPA's BASINS model is particularly useful, because it provides connections to GIS and several other ancillary resources for modeling.

HSPF is able to simulate the hydrologic and water quality processes on pervious and impervious land segments. It is also able to simulate chemical concentrations in one-dimensional and spatially lumped channels and storage confinements simulated as "well-mixed." The model uses continuous time-series rainfall and simulates all components of the hydrologic cycle to compute stream flow hydrographs and pollutographs. For water quantity calculations, HSPF accounts for infiltration, surface runoff, interflow, base flow, snowmelt, evapotranspiration, and ground-water interactions. For water quality, HSPF is able to simulate biochemical oxygen demand, temperature, pesticides, fecal coliforms, and sedimentation, as well as several specific chemicals including nitrogen, phosphorus, nitrite-nitrate, and plankton. The HSPF model uses hundreds of process algorithms developed from theory, laboratory experiments, and empirical relations based on data from numerous instrumented watersheds. These processes, which are grouped into modules, are explained in great detail in Bicknell et al. (1993).

5.3.3.2 Infiltration, Evaporation, and Sedimentation

HSPF's infiltration algorithms simulate both the time variation (based on soil moisture) and the spatial variation that occurs naturally over a pervious land segment (PERLND). These algorithms require a user defined initial infiltration rate and capacity, and also soil moisture and land slope information to determine the time-based infiltration fluctuations. A linear probability density function is used to account for the spatial variation of infiltration. These algorithms are based on the work of Philips (1957) and explained in detail in Bicknell et al (1993). Evaporation is simulated as time series of negative values and applied to each module.

HSPF simulates the production and removal of sediment through the SEDMNT section of the PERLND module, and the SINK subroutine within the channel and reservoir module (RCHRES). Detailed explanations of the complex algorithms used to calculate sedimentation are given by Donigian and Crawford (1976).

5.3.3.3 Modeling Decentralized Controls

Modeling BMPs in HSPF requires, at minimum, two modules. The RCHRES module simulates all processes that occur for a single reach of open channel or a completely mixed reservoir and may therefore be useful in modeling any BMP that will require storage. The PERLND module simulates components of the water budget to produce runoff quantities from a pervious land segment. This module may be useful in simulating BMPs that operate around infiltration. Using these two modules, HSPF is able to calculate total runoff volumes, peak flow rates, and generate hydrographs as an aid in modeling the structural BMPs listed in Table 5-4.

HSPF allows the user to model multiple modules in one simulation. Infiltration also allows different infiltration specifications for each PERLND module in the simulation. HSPF is also able to individually keep track of all other processes of the hydrologic cycle, including surface runoff, infiltration, evaporation, interflow and groundwater.

A toolkit designed specifically to model BMPs and LID was developed too late to be evaluated as part of this study. Specifically, the tool simulates BMPs that store stormwater runoff to attenuate flooding and remove pollutants. Presently, the tool is not recommended for simulating infiltration trenches, dry wells or BMPs that require channels that infiltrate. The toolkit is available from the U.S. EPA at: http://www.epa.gov/ATHENS/research/modeling/.

Limitations:

- Linear reservoir routing provides limited benefits for simulating storage-based BMPs.
- The comprehensive method of infiltration simulated in the PERLND module is not applied to the RCHRES module.
- Hydraulic routines are weak compared to SWMM, but water quality constituent interactions are stronger compared to SWMM.

Table 5-4. HSPF BMPs.

BMPs	HSPF
Ponds	Modeled in the "RCHRES" module within HSPF INFLOW: runoff from upstream source(s) (input of time series of precipitation and/or flow, solved using empirical algorithms) OUTFLOW: Solved using continuity equations LOSSES: ET and infiltration simulated by the input of a negative time series
Wetlands and bioretention facilities	Modeled in the "RCHRES" module within HSPF INFLOW: same as ponds OUTFLOW: same as ponds LOSSES: same as ponds

Table 5-4. HSPF BMPs (continued).

Infiltration trenches	Modeled in the "RCHRES" module OR Modeled in the "PERLND" module within HSPF **RCHRES module** INFLOW: same as ponds OUTFLOW: same as ponds LOSSES: same as ponds **PERLND module** INFLOW: Input of various time series OUTFLOW: Solved using the Manning equation and an empirical relationship of outflow depth to detention storage LOSSES: - Infiltration simulated by algorithms which represent time and spatial variability - Interception simulated by a user defined monthly capacity - ET calculated from potential evapotranspiration and an adjustment factor
Grass swales and filter strips	Modeled in the "RCHRES" module OR Modeled in the "PERLND" module within HSPF **RCHRES module** INFLOW: same as ponds OUTFLOW: same as ponds LOSSES: same as ponds **PERLND module** INFLOW: Input of various time series OUTFLOW: Solved using the Manning equation and an empirical relationship of outflow depth to detention storage LOSSES: - Infiltration simulated by algorithms which represent time and spatial variability - Interception simulated by a user defined monthly capacity - ET calculated from potential evapotranspiration and an adjustment factor

Table 5-4. HSPF BMPs (continued).

Dry wells	Modeled in the "RCHRES" module within HSPF INFLOW: same as ponds OUTFLOW: same as ponds LOSSES: same as ponds
Cisterns	Modeled in the "RCHRES" module within HSPF INFLOW: same as ponds OUTFLOW: same as ponds can specify an outflow schedule LOSSES: same as ponds
Porous Pavement	Modeled in the "PERLND" module within HSPF INFLOW: Input of various time series OUTFLOW: Solved using the Manning equation and an empirical relationship of outflow depth to detention storage LOSSES: - Infiltration simulated by algorithms which represent time and spatial variability - Interception simulated by a user defined monthly capacity - ET calculated from potential evapotranspiration and an adjustment factor

5.3.4 Western Washington Hydrology Model,Version 3 (WWHM3)

5.3.4.1 Model Overview

WWHM3 was created in 2001 by the Washington Department of Ecology, along with other private efforts. While based on the same algorithms that govern the HSPF model, WWHM3 adds an updated graphical user interface and capabilities targeted specifically at sizing stormwater facilities (Clear Creek Solutions, 2006). The latest version of WWHM3 is available at http://www.ecy.wa.gov/programs/wq/stormwater/wwhm_training/wwhm/wwhm_v3/index.html).

WWHM3 and HSPF calculate runoff from a basin using the same framework (PERLND and/or IMPLND modules). However, WWHM3 has calibrated the runoff process algorithms to 19 counties in western Washington.

WWHM3 is worthy of discussion because it explicitly addresses most common BMPs, while using the proven HSPF computational engine. The program has specific elements for several different storage units as well as elements that would work well for infiltration-based BMPs. Infiltration is simulated within the program based as an actual measured value and evaporation data are required in a time series format.

Also, the user has the option of selecting the U.S. EPA SWMM routing algorithms for use in place of HSPF's RCHRES routing. The advantage to using SWMM routing is that it has the ability to implement the dynamic wave method, which solves the complete one-dimensional Saint Venant flow equations, thusly theoretically producing the most accurate results. Dynamic wave routing can account for channel storage and pressurized flow, which HSPF's linear reservoir routing cannot (Beyerlein, 2005).

5.3.4.2 Infiltration, Evaporation, and Sedimentation

Infiltration is simulated within the program based on an actual measured value and a reduction factor. The infiltration reduction factor ranges from 0-1 and is user specified, acting as an inverse factor of safety. Infiltration is simulated in each element. The user has the option of simulating infiltration through the entire wetted surface or the bottom of the facility only. Evaporation is simulated using a time series of negative values and applied to each element. WWHM3 does not address sedimentation effects on BMPs.

5.3.4.3 Modeling Decentralized Controls

Like HSPF, WWHM3 can individually keep track of all processes of the hydrologic cycle. WWHM3 can then route these water quantities through a variety of different elements (e.g. pond, channel, gravel trench bed elements) to design the seven structural BMPs listed in Table 5-5. Because WWHM3 has many specific elements with which to model BMPs, the need for routing runoff from one land segment to another is eliminated. Using an AutoPond feature, WWHM3 can even automatically size ponds depending upon the user-specified permanent pool depth and amount of freeboard.

WWHM3 can also explicitly simulate the effects of some decentralized controls on stormwater (simulated as runoff plus interflow in the model), groundwater and evaporation quantities using an LID generator. The decentralized practices simulated are:

- Roof runoff dispersion and infiltration on an adjacent basin element
- Green roofs

The user is able to select any combination of the decentralized controls, apply them to basin elements, and observe stormwater changes between different combinations. This allows the user to pick the best possible combination of decentralized practices.

Limitations:

- While different infiltration is allowed among BMPs, it is held at a constant value (does not change with time).
- Runoff processes would need new calibration to apply to regions outside of western Washington.
- Sedimentation is not simulated.

Table 5-5. WWHM3 BMPs.

BMPs	WWHM3
Ponds	Modeled as a "Pond element" within WWHM3 INFLOW: runoff from upstream source(s) (input of time series of precipitation and/or flow) OUTFLOW: - stage-storage-discharge relationship calculated once pond dimensions and orifices are entered - multiple outlets allowed: up to 3 orifices + riser/weir LOSSES: - infiltration simulated based on a measured infiltration rate multiplied by a reduction factor (which acts as a factor of safety) - evaporation simulated by negative time series

Table 5-5. WWHM3 BMPs (continued).

<table>
<tr><td rowspan="2">Wetlands and bioretention facilities</td><td colspan="2">Modeled as a "Pond element"
OR
Modeled as a "Channel element" within WWHM3</td></tr>
<tr><td>Pond element
INFLOW:
same as ponds

OUTFLOW:
same as ponds

LOSSES:
same as ponds</td><td>Channel element
INFLOW:
same as ponds

OUTFLOW:
solved using Manning's equation

LOSSES:
- infiltration simulated in same way as for ponds
- ET not simulated within the channel</td></tr>
<tr><td>Infiltration trenches</td><td colspan="2">Modeled as a "Gravel Trench Bed" element within WWHM3

INFLOW:
same as ponds

OUTFLOW:
same as ponds (can have one or more surface outlets if desired)

LOSSES:
same as ponds</td></tr>
<tr><td rowspan="2">Grass swales and filter strips</td><td colspan="2">Modeled as a "Pond element"
OR
Modeled as a "Sand Filter" element within WWHM3</td></tr>
<tr><td>Pond element
INFLOW:
same as ponds

OUTFLOW:
same as ponds

LOSSES:
same as ponds</td><td>Sand Filter element
INFLOW:
same as ponds

OUTFLOW:
solved using Darcy's Law (Q=KIA)

LOSSES:
same as ponds</td></tr>
</table>

Table 5-5. WWHM3 BMPs (continued).

<table>
<tr><td>Dry wells</td><td>Modeled as a "Tank" element (box shaped)
OR
Modeled as a "Vault" element within WWHM3 (cylinder shaped)

Tank element
INFLOW:
same as ponds

OUTFLOW:
same as ponds

LOSSES:
same as ponds

Vault element
INFLOW:
same as ponds

OUTFLOW:
same as ponds

LOSSES:
same as ponds</td></tr>
<tr><td>Cisterns</td><td>Modeled as a "Tank" element (box shaped)
OR
Modeled as a "Vault" element within WWHM3 (cylinder shaped)

Tank element
INFLOW:
same as ponds

OUTFLOW:
same as ponds

LOSSES:
same as ponds

Vault element
INFLOW:
same as ponds

OUTFLOW:
same as ponds

LOSSES:
same as ponds</td></tr>
<tr><td>Porous Pavement</td><td>Modeled as a "Basin" element
OR
Modeled as a "Gravel Trench Bed" element within WWHM3

Basin element
INFLOW:
input of time series of precip and/or flow

OUTFLOW:

solved using empirical water balance

LOSSES:
same as ponds

Gravel Trench Bed element
INFLOW:
same as ponds

OUTFLOW:
same as ponds (can have one or more surface outlets if desired)

LOSSES:
same as ponds</td></tr>
</table>

5.4 Conclusions

This chapter presents these four models in the context of modeling decentralized controls and provides general conclusions as to the practicality of their use for decentralized controls.

The U.S. EPA SWMM's long history and capability has secured it as a popular choice in BMP design. Even though no "formal" support is offered by the U.S. EPA, there is no shortage of support amongst professionals via organized e-mail lists (e.g., SWMM-USERS@listserv.uoguelph.ca), internet forums, and very comprehensive help files and users manuals. SWMM's inherent ability to route flow from an impervious surface to a pervious surface (and from one subcatchment to another) makes it an ideal candidate for simulating decentralized controls. It is also the most comprehensive regarding the simulation of infiltration, as it applies several well known techniques and allows differentiation between BMPs. SWMM's success with decentralized controls has been documented in detail in Huber et al. (2006).

TR-55 is perhaps the least capable of the four in this report for modeling decentralized controls. The most crucial unit processes, infiltration and evaporation, are consolidated into the initial abstraction term, and ultimately put a large emphasis on the curve number. The SCS equations in their current form do not allow any degree of differentiation between the two that would be necessary for infiltration trenches, porous pavements, filter strips, or other BMPs that exhibit high infiltration and low evaporation (or vice versa). Furthermore, the model does not account for these losses in structural reaches. Doing so would require further analysis outside of the software with the model output, subject to the judgment of the user. If one were to use this model for simulating a decentralized control system, the setup would likely require multiple TR-55 simulation files, simply because one file would lack the routing complexity of a multiple BMP decentralized system. These reasons suggest that TR-55 is not truly intended for BMP design beyond basic detention volumes, and therefore seems more suited for centralized control design on a macro scale. Its strengths instead lie in its simplicity to generate hydrographs with little time and input. It appears that the use of TR-55 in simulating decentralized controls is in the early stages of the planning process, not in actual BMP modeling. Watershed planners may find it viable to get a broad, general idea of a watershed's developmental response and/or problem areas based solely on water quantities.

The HSPF model is geared toward both water quality and water quantity analyses. HSPF is most commonly used not as a BMP design package, but instead as a tool to assess the water quality effects of BMPs as well as land-use changes on a watershed scale. Its strengths lie in its ability to account for the fate of multiple contaminants during point and non-point pollution sources through continuous simulation as well as its incorporation into U.S. EPA's comprehensive BASINS modeling suite. It is able to produce peak flows, volumes and hydrographs, and account for the vertical water balance in ways that are very useful for simulation of distributed controls. HSPF is also supported by a list-server: HSPF-USERS@listserv.uoguelph.ca.

WWHM3, although very new, seems to be a very promising addition to the crowded field of hydrologic models. The model is proving to be a useful and diverse tool for many private consulting firms across western Washington. The model's flexibility has allowed it to be used for hydraulic purposes outside of its original design, as it has been applied to culvert sizing and wetland hydro-period analyses. It possesses many qualities that make it a viable decentralized controls choice: elements that explicitly model many BMPs, many land segments allowed per simulation, and loss rates that can be assigned differently for each element. However, the process algorithms need calibration for simulations outside of western Washington.

Designating which model is more suitable for simulating decentralized controls is difficult without actual data testing. Such testing, however, was not within the scope of this report. A likely test would include a comparison of each models behavior in an aggregated vs. disaggregated simulation, singling out the model with the least margin of error. It would also be beneficial to see, using the same data set, the computed reduction in runoff each model produces when implementing the same decentralized controls. Comparing the results of these two tests would provide much more insight into the limitations of each model.

APPENDIX A

SWMM5 BMP TEMPLATES

A.1 Introduction

This collection of templates was created using the U.S. EPA's Storm Water Management Model Version 5 (SWMM5). The model, documentation, and source code are all available at no charge from the U.S. EPA at http://www.epa.gov/ednnrmrl/models/swmm/index.htm. The intended use of these templates is to provide a framework from which the user can begin to use SWMM5 to evaluate decentralized controls. A template exists for each of the following seven representative structural BMPs:

1. Ponds
2. Wetlands
3. Infiltration trenches
4. Grass swales and filter strips
5. Dry wells
6. Cisterns
7. Porous pavements

Ponds and wetlands are included among these BMPs even though the emphasis of this report is on decentralized, low impact development-like controls because storage is an inherent unit process of most controls and ponds exemplify how this is done in SWMM, and wetlands may be thought of as large versions of bioswales and similar ponded water devices in which vegetation plays a key role. Additional decentralized controls, such as bioretention and green roofs, may be simulated in SWMM, perhaps with some ingenuity, similarly to grass swales and filter strips. Huber et al. (2006) provide insights into using SWMM and other models for a variety of conventional BMPs and distributed wet-weather controls.

Each template described below is set up to run upon opening of the file under two different precipitation conditions. This way, the user is able to see the structure of a running SWMM5 model, and then make alterations to the template in order to adapt it to his/her project. These alterations include, but are not limited to, changes in the physical characteristics of the subcatchment, physical characteristics of the BMP itself, and the storm events.

Throughout this documentation, references will be made to certain menus, toolbars and areas of SWMM5's graphical user interface (GUI). Figure A-1 below shows SWMM5's GUI.

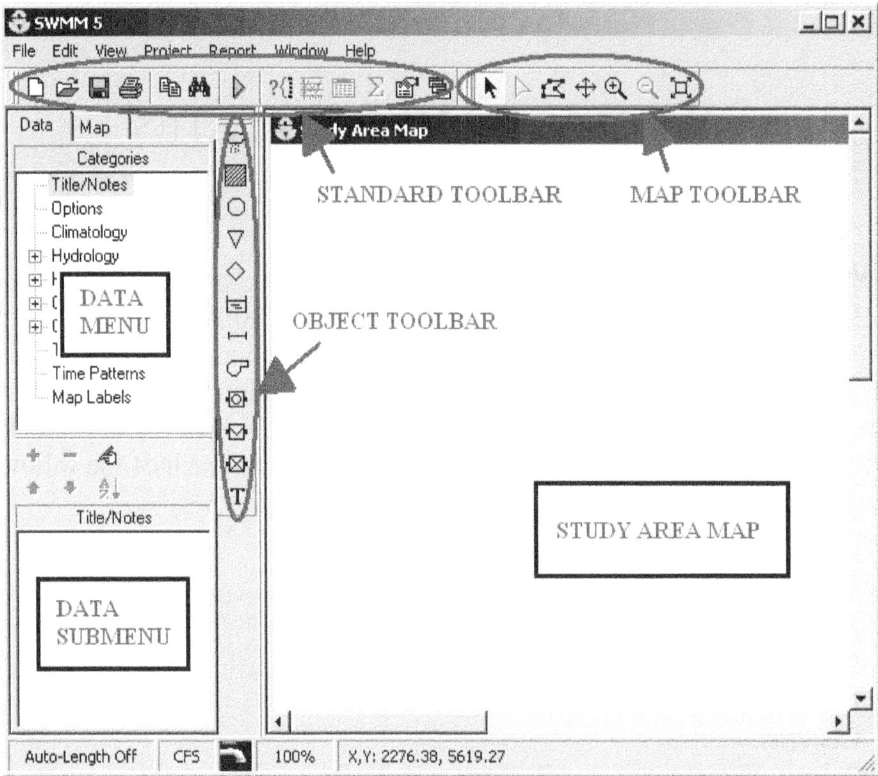

Figure A-1. SWMM5 GUI.

To demonstrate some of SWMM's capabilities, and also the diversity in design approaches, each template is complete with two different storm events to simulate. Upon initial loading into SWMM, the templates are defaulted to run with precipitation input from the rain gage titled "Storm." This scenario simulates a short duration, high-intensity storm of 3 inches (total volume) of precipitation in only 6 hours. The time series attached to this rain gage varies the rainfall intensity every hour for the 6-hour duration.

The second precipitation simulation is attached to the rain gage titled "WaterYear." This scenario simulates 15-minute rainfall depths (instead of hourly intensities used in "Storm") over the course of an entire water year. The water year used is that of Colton, Oregon, for 1996, as obtained from National Oceanic and Atmospheric Administrations's National Climatic Data Center. Only one rain gage can be attached to a subcatchment at any one time (but different subcatchments may access different gages).

The method for changing simulations is as follows:

1. Select the Select Object button from the Map Toolbar.
2. Double click on Subcatchment1 in the Study Area Map to bring up the Properties window.
3. Click on the Rain Gage pull down menu within the Properties window and select WaterYear. This will reference the WaterYear rain gage instead of the Storm rain gage.
4. Next, click on Options in the Data Menu
5. In the Data Submenu, double click Dates to bring up the Dates tab of the Simulation Options window.
6. Change the date in the "End Analysis on" field to 09/30/1996 and click OK.

Following the steps above will allow SWMM to run and produce graphical and tabular output for an entire water year.

When possible, using rainfall data with the smallest time interval is ideal. One should also use an appropriate simulation time step according to the size of the simulation. Simulation time steps should be short for small urban runoff simulations (on the order of 1 minute for rooftops) and perhaps as long as 5 minutes for larger urban subcatchments of tens of acres. Doing so allows the user to monitor more closely the behavior of the BMP. Rainfall intensities often vary considerably over time. High rainfall intensities over short periods can have a significant effect on peak flows and flood frequencies. The National Climatic Data Center has rainfall data available down to the 15-minute level for many major cities in the United States. However, the resolution for these 15-minute data is usually 0.1 in, instead of a more ideal and standard 0.01 in. When small-time-step data are not available, various rainfall disaggregation techniques (such as those outlined in Guntner et al., 2001) have been employed with various levels of success.

A.2 SWMM5 BMP Templates

A.2.1 Ponds

Template filename(s): "Pond – Storage Unit.INP"

A pond is a device that stores and discharges runoff at a controlled rate. Detention ponds are usually employed as a means to collect runoff from an area of new imperviousness, store it for a period of time, and discharge back into an existing system at a rate similar to the pre-existing condition.

Ponds are modeled as a Storage Unit node in SWMM. There are three common types of ponds, all of which can be modeled in SWMM using this template:

1. Retention pond: a pond with a continuous pool of water
2. Detention pond: a pond that has water in it for only a short period of time following a precipitation event
3. Extended dry detention pond: a pond that drains slower than a detention pond

Because these types of ponds are so similar, differentiating between the three in SWMM only involves manipulation of the outlet(s). A retention pond, for example, will have its outlet higher than the invert elevation of the pond, depending on the permanent pool depth desired. An extended dry detention pond, on the other hand, may just have a significantly smaller orifice at the pond invert than a regular detention pond to regulate flow to a slower rate (Figure A-2). Ponds can be designed to provide biological and sedimentation treatment.

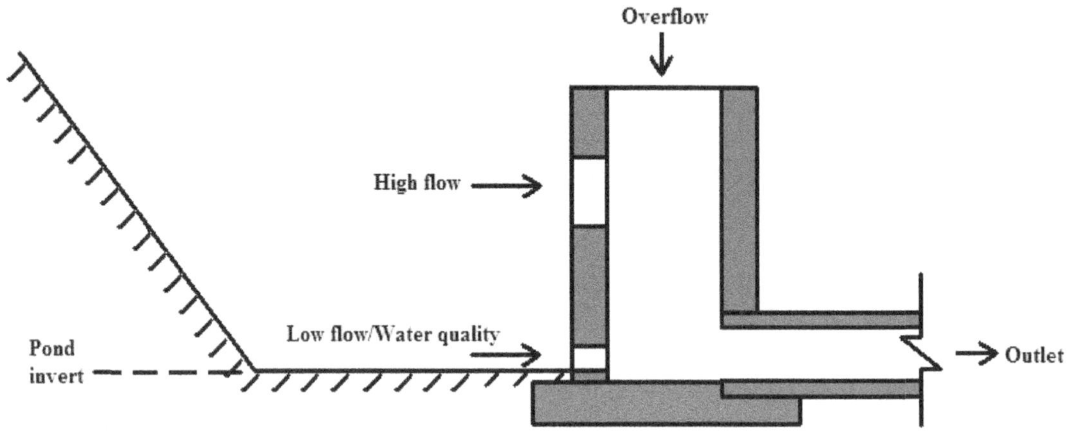

Figure A-2. Typical Pond Outlet Structure.

A.2.1.1 Inflow

The inflow to the pond is from runoff from Subcatchment1. Within this template, Subcatchment1 has an area of 3 acres, slope of 0.5%, width of 400 feet, and impervious area percentage of 50%. Typical applications of Subcatchment1 can be a small, medium-density urban residential development, a paved parking lot, or similar land area of equal imperviousness. Runoff from Subcatchment1 is routed to Junction1, through Conduit1 (a 100-foot long circular pipe, with 2% slope), and finally to the Pond node itself. The outlet of Conduit1 is offset upward of the pond invert elevation by 3 feet to prevent surcharging (Figure A-3).

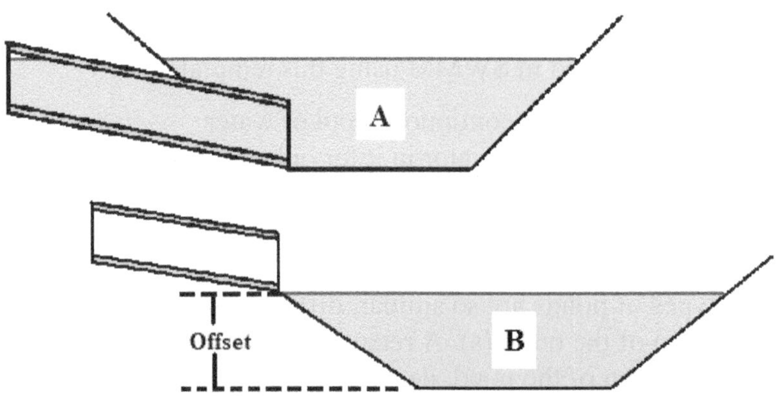

Figure A-3. (A) Conduit Invert is Placed at Invert of Pond.
(B) Conduit invert is Offset Above Pond Invert, Preventing Surcharging.

Changes to the specifics of Subcatchment1 can be made by selecting the Select Object button from the Map Toolbar, then double clicking on Subcatchment1 in the Study Area Map. This will bring up the properties window for Subcatchment1, in which the parameters discussed above may be altered.

A.2.1.2 Pond Specifics

The sizing of the pond in this template was determined using the most intense of the two precipitation events. The pond is rectangular in shape (30 feet by 15 feet) and has side slopes of 3:1 (horizontal:vertical), which are common parameters in pond design (Clean Water Services, 2004). Entering pond geometry into SWMM requires the use of a Storage Curve, which relates the depth of the pond to its surface area. The Storage Curve loaded with this template is shown in Figure A-4 below.

	Depth (ft)	Area (ft2)
1	102	450
2	103	756
3	104	1134
4	105	1584
5		

Figure A-4. The Storage Curve.

To change the Storage Curve for the pond, click on Curves→Storage Curves in the Data Menu, then double click the Pond storage curve in the Data Submenu to bring up the Storage Curve editor. Change the values in the fields as necessary, then click OK.

A.2.1.3 Outflow

Outflow from the pond is controlled by three devices in this template, but even more may be applied. Two circular orifices and a weir (each at different heights) are present in the template. The invert of Orifice1 is located at the invert elevation of the pond, Orifice2 is 1 foot above, and the weir elevation is 2.5 feet above the pond invert elevation. Both orifices are circular in shape. Orifice1 and Orifice2 measure 4 and 6 inches in diameter, respectively. The weir is to be considered as the pond overflow device (only active if the pond nears or exceeds its capacity under demanding storm conditions). Another orifice may be applied in place of a weir in this same manner if the user wishes.

Alterations to the size, shape, height, and many other parameters of each orifice/weir can be made by clicking on the Select Object button in the Map Toolbar and then double clicking on the orifice or weir in the Study Area Map.

A.2.1.4 Losses

Evaporation and infiltration are simulated in the pond. Evaporation can be entered as a constant value, as a time series, as a climate file, or as monthly average values.

For this template, average daily evaporation rates for each month were obtained for Corvallis, Oregon, from the U.S. Bureau of Reclamation's AgriMet website (http://www.usbr.gov/pn/agrimet/). These values are shown in Figure A-5.

Monthly Evaporation (in/day)

Jan	Feb	Mar	Apr	May	Jun
0.02	0.04	0.07	0.12	0.17	0.22
Jul	**Aug**	**Sep**	**Oct**	**Nov**	**Dec**
0.28	0.25	0.18	0.08	0.03	0.02

Figure A-5. Daily Average Evaporation Rates for Each Month.

To change these values, or select another method of simulating evaporation, click on Climatology in the Data Menu. Next, double click on Evaporation in the Data Submenu to bring up the Evaporation tab of the Climatology Editor. Change the values in the fields as necessary, then click OK. To apply these evaporation values to the pond, the Evaporation Factor must be changed in the pond's Properties window. To access the properties window, select the Select Object button from the Map Toolbar and double click on the Pond Storage Unit node. The evaporation factor must be changed to a value ranging from zero (to simulate zero evaporation) to one (to represent full evaporation).

SWMM does not explicitly address infiltration that occurs in Storage Units. However, infiltration may be directly applied to any node by incorporating the use of a negative time series of flow values. Alternatively, one may also add the values of infiltration to evaporation and change the evaporation values in the Climatology Editor accordingly. This template uses the earlier method.

An infiltration rate of 0.5 inch per hour was used. When applied to the bottom dimensions of the pond (20 feet by 10 feet), this produces a value of 0.002 cubic feet per second (cfs). A separate time series was created for infiltration, and this value is kept constant. Changes to the infiltration values can be made by first clicking on Time Series in the Data Menu, then double clicking on Infiltration in the Data Submenu to bring up the Time Series Editor window. Change the values in the fields as necessary, then click OK.

This time series of infiltration is applied to the pond through the Properties window. Highlighting the Inflows field and clicking on the ellipsis button brings up the Inflows for Node window. In this window under the Direct tab, the Infiltration time series is selected under the Time Series field. This applies the infiltration values to the pond node.

A.2.2 Wetlands

Template filename(s): "Wetland – Conduit.INP"
"Wetland – Storage Unit.INP"

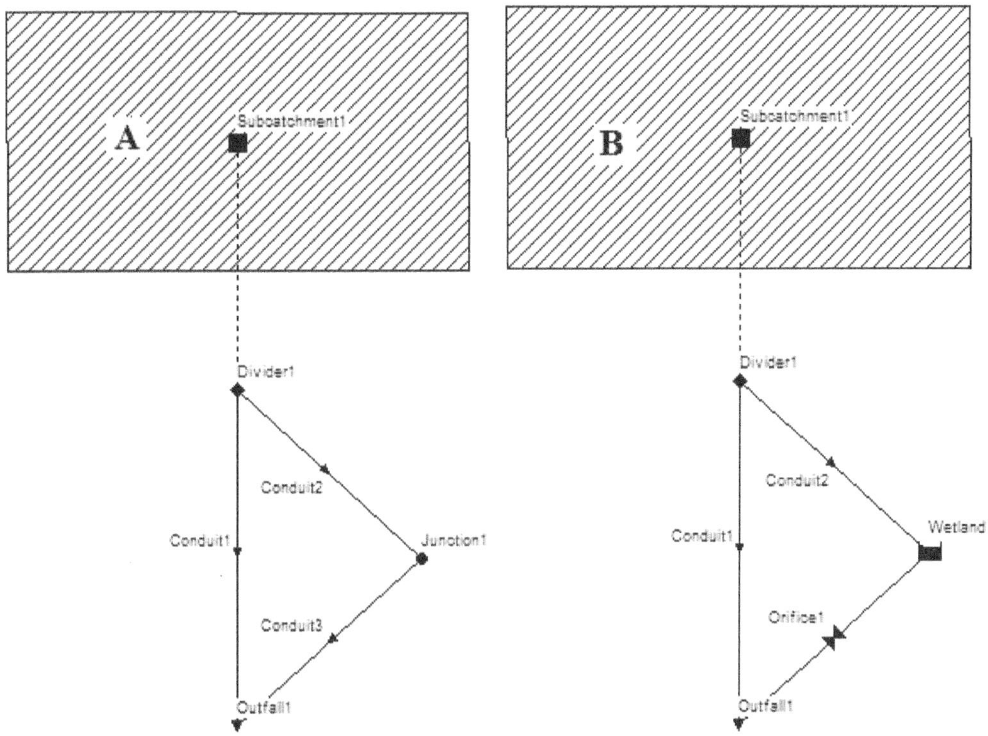

Figure A-6. (A) Plan View of the "Wetland – Conduit.INP" Template. (B) Plan View of the "Wetland – Storage Unit.INP" Template.

Constructed wetlands are defined as areas of land that become inundated or saturated as a result of the rising of ground and/or surface water levels (Figure A-6). Wetlands are found adjacent to streams and have dense vegetation, limiting water velocities to very low levels. Wetlands are home to various plant and animal life and can be designed to provide controlled discharge and biological and sedimentation treatment.

Wetlands can be modeled as either a series of Conduits or as a Storage Unit node in SWMM. A series of Conduits may be more applicable if the wetlands will indeed behave as more of a channel (that is, velocities are involved and Manning's n is deemed an important design factor). However, a Storage Unit allows the user to much more accurately assign sizing, shape, and water release characteristics of the wetlands.

A.2.2.1 Inflow

The inflow to the wetlands in both templates is provided by the runoff induced by Subcatchment1. Subcatchment1 has an area of 300 acres, slope of 1%, width of 2000 feet, and impervious area percentage of 50%. A typical application of Subcatchment1 is a small creek drainage basin in an urban setting. Runoff from Subcatchment1 is routed first to Divider1, where flow is divided to either Conduit1 or to the wetlands system itself. The systems in place here – either the wetland channel or wetland conduit – represent the wetland inundation that would occur given high creek or river flows (represented by Conduit1).

SWMM accepts four ways in which to divert flow to the wetlands from Divider1. These methods are:

1. Cutoff: after a certain flow is realized in Conduit1, everything above that flow is diverted to the wetlands
2. Tabular: incorporates the use of a table to relate diverted flow to total node inflow
3. Weir: uses a weir set at a specific elevation about the divider invert elevation to calculate the divided flow
4. Overflow: diverts all water above the capacity of Conduit1

These templates use Overflow as the diversion type. A constant initial value of 105 cfs is also applied to the Diversion1 node to simulate a baseflow element.

Changes to the specifics of Subcatchment1 or Divider1 can be made by selecting the Select Object button from the Map Toolbar, then double clicking on the item in the Study Area Map. This brings up the properties window for each item, in which all parameters can be altered. Changes to the baseflow applied to Divider1 is done by opening the Properties window for the node, highlighting the Inflows value and clicking the ellipsis button, clicking on the Dry Weather tab in the Inflows for Node window, and changing the value in the Average Value field.

A.2.2.2 Wetland Specifics

For the "Wetland – Conduit.INP" template, Conduit2 and Conduit3 represent the wetlands. Both conduits together total 600 feet long, are trapezoidal, and have a Manning's *n* value of 0.075. The conduits are 30 feet wide at their invert elevation, have a maximum depth of 2 feet, and have 3:1 side slopes. SWMM supports all manner of channel geometries. Changing amongst them is done by opening the Properties window of the conduit, selecting the Shape field, and clicking the ellipsis button to bring up the Cross-Section Editor window.

For the "Wetland – Storage Unit.INP" template, the wetlands is represented by the Storage Unit titled "Wetlands.." The sizing of the Storage Unit represents a fairly large area adjacent to the stream to be used as the wetlands. The wetlands are depicted as a rectangular-shaped Storage Unit measuring 450 feet by 50 feet, with 3:1 side slopes (horizontal:vertical). Entering Storage Unit geometry into SWMM requires the use of a Storage Curve, which relates the depth of the Storage Unit to its surface area. The Storage Curve loaded with this template is shown in Figure A-7.

	Depth (ft)	Area (ft2)
1	100.1	18000
2	101.1	20976
3	102.0	24024
4		

Figure A-7. The Storage Curve Included in the "Wetlands – Storage Unit.INP" Template.

Changes to the Storage Curve for the wetland are performed in the manner discussed in Section A.2.1.2.

A.2.2.3 Outflow

Note that since conduits are used in the "Wetland – Conduit.INP" template, no storage is provided. That is, inflow and outflow to and from the wetlands are equal and no outflow control is provided.

For the "Wetland – Storage Unit.INP" template, flow exiting the Wetland Storage Unit node is controlled by Orifice1 (as seen in Figure A-6). The orifice is rectangular in shape to more accurately depict a small side channel that may exist in its place to "drain" the wetlands back into the main channel. The orifice is 10 ft wide and shares the same maximum depth as the wetland.

Alterations to the orifice are performed in the manner presented in Section A.2.1.3.

A.2.2.4 Losses

Both evaporation and infiltration are simulated in the "Wetland – Conduit.INP" template. However, SWMM does not explicitly address these two losses that can occur within Conduits. Both of these losses are applied by creating a time series of negative flows and applying it to Junction1. Junction1 exists solely for this purpose. In order for this to work, infiltration and evaporation must be combined into one time series, and the values within the time series must be expressed in units of flow (CFS for these templates).

An infiltration rate of 0.5 in./hr was used. When applied to the bottom dimensions of the entire wetland system (both Conduit2 and Conduit3, 600 feet by 40 feet total), this produces a value of 0.278 cfs. Average daily evaporation rates for each month were obtained for Corvallis, Oregon, from the U.S. Bureau of Reclamation's AgriMet website. These values are shown in Figure A-5.

The time series titled "Losses" combines the constant value of infiltration and the evaporation rates for each month. This time series is applied to Junction1 through the Properties window. Highlighting the Inflows field in the properties window and clicking on the ellipsis button brings up the Inflows for Node window. In this window under the Direct tab, the Losses time series is selected under the Time Series field. This applies infiltration and evaporation to Junction1, and the wetland system itself.

Evaporation and infiltration are also simulated in the "Wetland – Storage Unit.INP" template. Evaporation can be entered as a constant value, as a time series, as a climate file, or as monthly average values. AgriMet's Corvallis, Oregon, values shown in Figure A.5 were used. Changes to the evaporation values are performed as presented in Section A.2.1.4.

SWMM does not explicitly address infiltration that occurs in Storage Units. However, infiltration is applied to the Wetland node by incorporating the use of a negative time series of flow values. Alternatively, one may also add the values of infiltration to evaporation and change the evaporation values in the Climatology Editor accordingly. This template uses the prior value.

An infiltration rate of 0.5 inch per hour was used. When applied to the bottom dimensions of the wetland (450 feet by 50 feet), this produces a value of 0.260 cfs. A separate time series was created for infiltration, and this value is kept constant. Changes to the infiltration values are performed as presented in Section A.2.1.4.

This time series of infiltration is applied to the Wetland Storage Unit node through the Properties window. Highlighting the Inflows field and clicking on the ellipsis button brings up the Inflows for Node window. In this window under the Direct tab, the Infiltration time series is selected under the Time Series field. This applies the infiltration values to the Wetland node.

A.2.3 Infiltration Trenches

Template filename(s): "Infiltration Trench – Storage Unit.INP"
"Infiltration Trench – Subcatchment.INP"

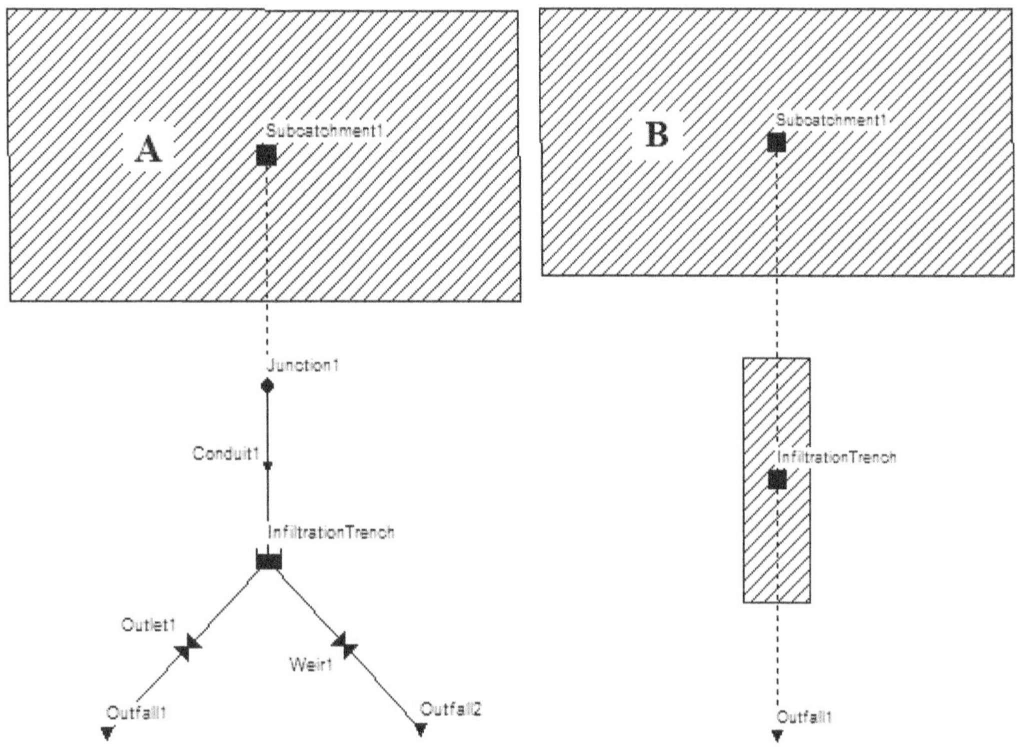

Figure A-8. (A) Plan View of the "Infiltration Trench – Storage Unit.INP" Template; (B) Plan View of the "Infiltration Trench – Subcatchment.INP" Template.

An infiltration trench is a rock filled trench with vertical sides, designed to receive stormwater runoff from small impervious areas (typically no greater than 5 acres) (Huber et al., 2006). Infiltration trenches are usually designed to be long and narrow, and are filled with a gravely soil of high void ratio. Because sedimentation is such a major cause of infiltration trench failure, proper design measure usually includes the conveyance of the runoff through a pretreatment device before entering the trench (i.e., swale or similar) (Figure A-8). Runoff then enters the trench and is stored in the void space until infiltrated into the soil through the bottom and sides of the trench. Infiltration trenches typically have no outlet, but some jurisdictions mandate an overflow failsafe of some kind. Sizing of the trench is done in the same manner that any storage device is sized. Usually one or more design storms are enlisted and the trench must handle the entire volume of the storm, and nothing more, without overflowing. SWMM's continuous simulation capability provides the better option for determining a design volume that will overflow only with a specified frequency.

Infiltration trenches can be modeled as either a Storage Unit node or as a Subcatchment in SWMM. When using a Storage Unit, the user has greater control over the geometry and overflow discharge characteristics of the infiltration trench. Subcatchments, on the other hand, incorporate the use of more accurate infiltration estimation methods (Horton, Green-Ampt, or Curve Number methods).

A.2.3.1 Inflow

The inflow to the infiltration trenches in both templates is provided by the runoff induced by Subcatchment1. Subcatchment1 has an area of 3 acres, slope of 0.5%, width of 500 feet, and impervious area percentage of 50%. Typical applications of Subcatchment1 can be a small, medium-density urban residential development, a paved parking lot, or similar land area of equal imperviousness.

For the "Infiltration Trench – Storage Unit.INP" template, runoff from Subcatchment1 is routed to Junction1 and through Conduit1 before reaching the infiltration trench. Conduit1 represents the pretreatment and conveyance swale. Conduit1 is a typical trapezoidal open channel, 100 feet long, with 1% slope and 3:1 side slopes (horizontal:vertical). The outlet of Conduit1 is offset upward of the infiltration trench invert elevation by 4 feet to prevent surcharging.

For the "Infiltration Trench – Subcatchment.INP" template, runoff from Subcatchment1 is routed directly onto the infiltration trench. This is because of routing constraints in the SWMM program, which prevent the routing from a subcatchment to a conduit and then to another subcatchment.

Changes to the specifics of Subcatchment1 and Conduit1 are performed in the manner discussed in Section A.2.1.1

A.2.3.2 Infiltration Trench Specifics

For the "Infiltration Trench – Storage Unit.INP" template, the infiltration trench is represented by the Storage Unit titled "InfiltrationTrench." The trench is rectangularly shaped and measures 60 feet long, 5 feet wide, and 4 feet deep with vertical sides. Entering Storage Unit geometry into SWMM requires the use of a Storage Curve, which relates the depth of the Storage Unit to its surface area. The Storage Curve loaded with this template is shown in Figure A-9.

	Depth (ft)	Area (ft2)
1	100	300
2	101	300
3	102	300
4	103	300
5	104	300
6	105	300

Figure A-9. The Storage Curve Included in the "Infiltration – Storage Unit.INP" Template.

Changes to the Storage Curve for the infiltration trench are performed in the manner discussed in Section A.2.1.2.

For the "Infiltration Trench – Subcatchment.INP" template, the infiltration trench is represented by the subcatchment titled "InfiltrationTrench." This trench has a rectangular shape, measuring 50 feet long, 5 feet wide, and 4 feet deep with vertical sides. SWMM uses several fields in the Properties Window to simulate this geometry for a subcatchment. The value in for the Area property corresponds to the surface area of the infiltration trench (50 feet by 5 feet) and is expressed in units of acres, producing a value of 0.006 acre. A value of 5 feet is used in for the Width property. To reflect the depth of the infiltration trench, the value for the Dstore-Perv property (depth of depression storage on the pervious portion of the subcatchment) is the depth of the trench in inches (48 inches). The subcatchment is represented as 100% pervious, so the value in the %Imperv property is 0. Very low slope and high Manning's n properties are used for the trench to minimize outflow.

Changes to the specifics of the infiltration trench subcatchment are performed in the manner discussed in Section A.2.1.1.

A.2.3.3 Outflow

When designing an infiltration trench, outflow from the trench should be kept to a minimum within the model. This is, in fact, an infiltration trench's primary function, to store and slowly infiltrate what would otherwise be runoff. This is done in the model by utilizing the properties described above (low slope, high Manning's *n* value, etc.). However, the designer should always know and account for the pathway of water that might overflow during events that exceed the design. Depending on the design criteria, an outflow into an existing system may be desired due to trench sizing constraints for example. Both templates incorporate an overflow.

For the "Infiltration Trench – Storage Unit.INP" template, Weir1 and Outfall2 represent the overflow outlet from the trench. The weir has an elevation of 4 ft above the invert elevation of the trench and a width of 5 feet. Any flow that occurs over this weir does so when the trench has exceeded its capacity. Edits to these parameters is done in the Properties window of Weir1.

For the "Infiltration Trench – Subcatchment.INP" template, overflow from the infiltration trench is routed directly to Outfall1. Any flow that occurs to the node Outfall1, does so as runoff from the infiltration trench when the trench has reached its capacity.

A.2.3.4 Losses

Both evaporation and infiltration are simulated in the "Infiltration Trench – Storage Unit.INP" template. Evaporation can be entered as a constant value, as a time series, as a climate file, or as monthly average values.

For this template, average daily evaporation rates for each month were obtained for Corvallis, Oregon from the U.S. Bureau of Reclamation's AgriMet website. These values are shown in Figure A.5. Changes to the evaporation values are performed in the same manner discussed in Section A.2.1.4.

As discussed previously, SWMM does not explicitly address infiltration that occurs in Storage Units. If properly implemented, outflow rating curves can account for changes in the amount being infiltrated based on the water level in the storage unit. This template uses this last option.

An infiltration rate of 1.5 inches per hour was used. When applied to the bottom dimensions of the trench (50 feet by 5feet), this produces a value of 0.009 cfs. Infiltration begins to occur through the sides of the trench as the water level rises. Figure A-10 shows the rating curve created in SWMM to reflect this increase in infiltration. This rating curve is applied to Outlet1 in the template.

To change the Infiltration rating curve for the trench, click on Curves→Rating Curves in the Data Menu, then double click the Infiltration rating curve in the Data Submenu to bring up the Rating Curve editor. Change the values in the fields as necessary, then click OK.

	Head (ft)	Outflow (CFS)
1	0.0	0.009
2	0.5	0.011
3	1.0	0.014
4	1.5	0.020
5	2.0	0.028
6	2.5	0.037
7	3.0	0.049
8	3.5	0.062
9	4.0	0.077

Figure A-10. The Rating Curve included in the "Infiltration – Storage Unit.INP" Template to Reflect Infiltration Occurring as Depth Increases.

Evaporation and infiltration are also simulated in the "Infiltration Trench – Subcatchment.INP" template. Again, AgriMet's Corvallis, Oregon, values (shown in Figure A-5) were used for evaporation.

Infiltration occurring in the infiltration trench subcatchment is simulated in SWMM by using Horton infiltration model. The maximum and minimum infiltration rates are 3 inches per hour and 1.5 inches per hour, respectively. Changes to these parameters are made by opening the properties window for the infiltration trench subcatchment, highlighting the Infiltration field, and clicking on the ellipsis button. This brings up the Infiltration Editor window.

SWMM allows the use of one of three infiltration models (Horton, Green-Ampt, or Curve Number methods). Changing between the three requires opening the General tab in the Simulation Options window. Do this by first clicking on Options in the Data Menu, then double clicking on General in the Data Submenu.

A.2.4 Grass Swales and Filter Strips

Template filename(s): "Swale and Strip – Storage Unit and Subcatchment.INP"

Grass swales are earthen channels usually broad, shallow and trapezoidal in shape. These swales are designed to slowly convey, provide treatment, and infiltrate runoff from an adjacent road or parking lot. To a small degree, swales can also provide limited flood control.

Filter strips are flat, sloped, and vegetated overland flow paths that are intended to convey sheet flow from adjacent impervious areas. Filter strips are intended to filter runoff, slow water velocities and provide some level of treatment before the runoff enters another treatment or storage device. Typical applications of filter strips include placement directly next to a road way, near roof downspouts, and other small impervious areas.

These BMPs are modeled together in the same template. Grass swales are modeled as Storage Unit nodes and filter strips are modeled as Subcatchments within SWMM. The filter strip is placed upstream of the grass swale in this template.

A.2.4.1 Inflow

The inflow to the filter strip is provided by the runoff induced by Subcatchment1. Subcatchment1 has an area of 0.138 acre, slope of 2%, width of 100 feet, and is 100% impervious. Filter strip and grass swale combinations are commonly used adjacent to stretches of roadway or small parking lots. Subcatchment1 represents a 60 feet by 100 feet stretch of road or similar sized parking lot.

Runoff from Subcatchment1 is routed directly onto the filter strip. Runoff then infiltrates and evaporates, and the residual runoff is routed into the swale.

Changes to the specifics of Subcatchment1 are performed in the same manner discussed in Section A.2.1.1

A.2.4.2 Swale and Filter Strip Specifics

The filter strip is represented by the subcatchment titled "FilterStrip." The filter strip is rectangular shaped and runs the entire length of Subcatchment1. The filter strip measures 20 feet long (in the down slope direction), 100 feet wide, giving it a total area of 0.046 acre. The filter strip is completely pervious and has a slope of 4% (typical values of slope range from 2% to

15%). The filter strip is given a high Manning's *n* value of 0.24 to reduce velocities and reflect the vegetation that will be present. The value for the Dstore-Perv property (depth of depression storage on the pervious portion of the subcatchment) is 1 inch to reflect the small amount of storage that occurs on filter strips.

Changes to the specifics of the filter strip are performed in the same manner discussed in Section A.2.1.1.

The grass swale is represented by the storage unit node title "Swale." The swale is rectangular shaped and also shares the same length and Subcatchment1 and the filter strip. The swale measures 100 feet by 3 feet at its invert elevation, and side slopes are 4:1 (horizontal:vertical). Entering Storage Unit geometry into SWMM requires the use of a Storage Curve, which relates the depth of the Storage Unit to its surface area. The Storage Curve loaded with this template is shown in Figure A-11.

	Depth (ft)	Area (ft2)
1	100	300
2	101	1188
3	102	2204
4		

Figure A-11. The Storage Curve included in the "Swale and Strip – Storage Unit and Subcatchment.INP" Template.

Changes to the Storage Curve for the swale are performed in the same manner discussed in Section A.2.1.2.

A.2.4.3 Outflow

Outflow of the entire system is handled by a single weir (Weir1) connected to the swale. The weir has an elevation of 0.5 foot above the invert elevation of the swale and a width of 1 foot. This prevents surcharging and allows the swale to process the runoff as infiltration and/or evaporation. Edits to these parameters are done in the Properties window of Weir1.

A.2.4.4 Losses

Evaporation is simulated in both the filter strip and the swale. Evaporation can be entered as a constant value, as a time series, as a climate file, or as monthly average values.

For this template, average daily evaporation rates for each month were obtained for Corvallis, Oregon, from the U.S. Bureau of Reclamation's AgriMet website. These values are shown in Figure A-5. Changes to the evaporation values are performed in the same manner discussed in Section A.2.1.4.

Infiltration occurring in the filter strip subcatchment is simulated in SWMM by using Horton infiltration model. The maximum and minimum infiltration rates are 3 inches per hour and 0.5 inches per hour, respectively. Changes to these parameters are made by opening the properties window for the filter strip subcatchment, highlighting the Infiltration field, and clicking on the ellipsis button.

This template applies infiltration to the swale by the negative time series approach. An infiltration rate of 1inch per hour was used. When applied to the bottom dimensions of the swale (100 feet by 3 feet), this produces a value of 0.012 cfs. A separate time series was created for infiltration, and this value is kept constant. Changes to the infiltration values can be made by first clicking on Time Series in the Data Menu, then double clicking on Infiltration in the Data Submenu to bring up the Time Series Editor window. Change the values in the fields as necessary, then click OK.

This time series of infiltration is applied to the Swale Storage Unit node through the Properties window. Altering these infiltration parameters is performed in the same manner discussed in Section A.2.1.4.

A.2.5 Dry Wells

Template filename(s): "Drywell – Storage Unit.INP"

Dry wells are a basic variation of an infiltration trench or pond. Runoff is conveyed from building rooftops, driveways, or small parking lots to the dry well that lies beneath the ground surface. Therefore, the only loss mechanic that takes place in dry wells is infiltration. The use of drywells is beneficial when limited ground surface space is available.

Drywells are modeled in SWMM as a Storage Unit node.

A.2.5.1 Inflow

The inflow to the dry well in this template is provided by the runoff induced by Subcatchment1. Subcatchment1 has an area of 0.055 acre, slope of 50 %, width of 80 feet, and is 100 % impervious. In this template, Subcatchment1 represents a steep rooftop of a residential home or similar sized building.

Runoff from Subcatchment1 is routed to Junction1, through Conduit1, and then to the dry well itself. Conduit1 is a 50-foot long, circular pipe with a diameter of 1 foot and is given a very steep slope. Conduit1 represents a typical roof leader that is nearly vertical.

Changes to the specifics of Subcatchment1 and Conduit1 are performed in the same manner discussed in Section A.2.1.1.

A.2.5.2 Dry Well Specifics

The dry well is represented by the Storage Unit titled "DryWell." The well is cylindrical in shape to reflect the common use of concrete pipes as dry wells. The dry well is 10 feet in diameter and 5 feet tall. Entering Storage Unit geometry into SWMM requires the use of a Storage Curve, which relates the depth of the Storage Unit to its surface area. The Storage Curve loaded with this template is shown in Figure A-12 below.

	Depth (ft)	Area (ft2)
1	100	78.5
2	101	78.5
3	102	78.5
4	103	78.5
5	104	78.5
6	105	78.5
7		

Figure A-12. The Storage Curve included in the "Dry Well – Storage Unit.INP" Template.

Changes to the Storage Curve for the dry well are performed in the manner discussed in Section A.2.1.2.

A.2.5.3 Outflow

The dry well in this template is modeled with an overflow. Orifice1 and Outfall2 represent the overflow outlet for the dry well. The orifice is circular and has, diameter of 0.5 foot, and is placed at an elevation of 5 feet above the invert elevation of the dry well. Placing the orifice at this elevation means that any flow occurring through it does do when the well has exceeded its capacity. Edits to these parameters are done in the Properties window of Orifice1.

A.2.5.4 Losses

Since dry wells are placed underground, infiltration is the only loss simulated. SWMM does not explicitly address infiltration that occurs in Storage Units. If properly implemented, outflow rating curves can account for changes in the amount being infiltrated based on the water level in the storage unit. This template uses this last option.

An infiltration rate of 1.5 inches per hour was used. When applied to the bottom dimensions of the dry well (circular, 10 feet diameter), this produces a value of 0.003 cfs. Infiltration begins to occur through the sides of the trench as the water level rises. Figure A-13 shows the rating curve created in SWMM to reflect this increase in infiltration. This rating curve is applied to Outlet1 in the template.

Changes to the Infiltration rating curve for the well are performed in the manner discussed in Section A.2.3.4.

	Head (ft)	Outflow (CFS)
1	0.0	0.003
2	0.5	0.003
3	1.0	0.005
4	1.5	0.006
5	2.0	0.009
6	2.5	0.012
7	3.0	0.016
8	3.5	0.020
9	4.0	0.025

Figure A-13. The Rating Curve Included in the "Drywell – Storage Unit.INP" Template to Reflect Infiltration Occurring as Depth Increases.

Evaporation is only simulated on Subcatchment1. Evaporation can be entered as a constant value, as a time series, as a climate file, or as monthly average values.

For this template, average daily evaporation rates for each month were obtained for Corvallis, Oregon from the U.S. Bureau of Reclamation's AgriMet website. These values are shown in Figure A-5. Changes to the evaporation values are performed in the same manner discussed in Section A.2.1.4.

A.2.6 Cisterns

Template filename(s): "Cistern – Storage Unit.INP"

A cistern is any manner of storage device placed under a downspout for collection of roof drainage. The purpose of cisterns is usually not contaminant treatment related. Instead, they are designed to store water for later use and not promote infiltration and/or evaporation.

Cisterns are modeled in SWMM as a Storage Unit node. The modeling approach to cisterns is very similar to that of dry wells.

A.2.6.1 Inflow

The inflow to the cistern in this template is provided by the runoff induced by Subcatchment1. Subcatchment1 has an area of 0.17 acre, slope of 50%, width of 150 feet, and is 100 % impervious. In this template, Subcatchment1 represents a steep rooftop of a commercial or industrial building.

Runoff from Subcatchment1 is routed to Junction1, through Conduit1, and then to the cistern itself. Conduit1 is a 50-feet long, circular pipe with a diameter of 1 foot and is given a very steep slope. Conduit1 represents a typical roof leader that is nearly vertical.

Changes to the specifics of Subcatchment1 and Conduit1 are performed in the manner discussed in Section A.2.1.1.

A.2.6.2 Cistern Specifics

The cistern is represented by the Storage Unit titled "Cistern." The cistern is cylindrical in shape, with measurements of 10 feet in diameter and 10 feet tall. Entering Storage Unit geometry into SWMM requires the use of a Storage Curve, which relates the depth of the Storage Unit to its surface area. The Storage Curve loaded with this template is shown in Figure A-14.

Changes to the Storage Curve for the cistern are performed in the manner discussed in Section A.2.1.2.

	Depth (ft)	Area (ft2)
1	100	78.5
2	110	78.5
3		

Figure A-14. The Storage Curve Included in the "Cistern – Storage Unit.INP" Template.

A.2.6.3 Outflow

The cistern in this template is modeled with an overflow. Orifice1 and Outfall1 represent the overflow outlet for the dry well. The orifice is circular, has a diameter of 1 foot, and is placed at an elevation of 10 feet above the invert elevation of the dry well. Placing the orifice at this elevation means that any flow occurring through it does do when the cistern has exceeded its capacity. Edits to these parameters are done in the Properties window of Orifice1.

A.2.6.4 Losses

The water stored in cisterns may be used for a number of purposes. Common uses include outdoor washing, irrigation, and toilet flushing. Therefore, subtractions only occur from cisterns when needed. To model these subtractions, a water use schedule is need. Water use schedules can be represented in SWMM by a time series of negative flows.

The water use schedule included in this template is a basic example of an irrigation schedule. A 100-foot-by-100-foot field requires an inch of water per week for the months of March through October. This equates to about 119 cubic feet of water per week, or 0.001 cfs for this period. A separate time series was created for this water usage.

Changes to the water usage values can be made by first clicking on Time Series in the Data Menu, then double clicking on WaterSchedule in the Data Submenu to bring up the Time Series Editor window. Change the values in the fields as necessary, then click OK.

Evaporation is only simulated on Subcatchment1. Evaporation can be entered as a constant value, as a time series, as a climate file, or as monthly average values.

For this template, average daily evaporation rates for each month were obtained for Corvallis, Oregon from the U.S. Bureau of Reclamation's AgriMet website. These values are shown in Figure A-5. Changes to the evaporation values are performed in the same manner discussed in Section A.2.1.4.

A.2.7 Porous Pavement

Template filename(s): "Porous Pavement – Subcatchment.INP"

Porous pavements are a permeable pavement surface underlain by a stone reservoir. These pervious pavement surfaces infiltrate stormwater that would otherwise be runoff. The stone reservoir provides temporary storage for surface runoff before infiltrating into the soil. Because of their lower strength, porous pavements are typically used in areas of low traffic loading, such as parking lots and service roads.

Porous pavements are modeled in SWMM as subcatchments.

A.2.7.1 Inflow

The inflow to the porous pavement in this template is provided by the runoff induced by Subcatchment1. Subcatchment1 has an area of 0.17 acres, slope of 50%, width of 150 feet, and is 100% impervious. In this template, Subcatchment1 represents a steep rooftop of a commercial or industrial building. Runoff from Subcatchment1 is routed directly onto the porous pavement.

Changes to the specifics of Subcatchment1 are performed in the manner discussed in Section A.2.1.1.

A.2.7.2 Porous Pavement Specifics

The porous pavement is represented by the subcatchment titled "PorousPavement." The subcatchment is rectangular shaped, measures 100 feet long and 100 feet wide. SWMM uses several fields in the Properties Window to simulate this geometry for a subcatchment. The value in for the Area property corresponds to the surface area of the infiltration trench (100 feet by 100feet) and is expressed in units of acres, producing a value of 0.23 acres. A value of 100 ft is used in for the Width property. To reflect the depth of the total cross section of the porous pavement, the value for the Dstore-Perv property (depth of depression storage on the pervious portion of the subcatchment) is the depth of the pavement in inches (12 inches). The subcatchment is represented as 100% pervious, so the value in the %Imperv property is 0.

Changes to the specifics of the porous pavement subcatchment are performed in the manner discussed in Section A.2.1.1.

A.2.7.3 Outflow

Overflow from the porous pavement is routed directly to Outfall1. Any flow that occurs to this node does so as runoff from the porous pavement when it has reached its capacity.

A.2.7.4 Losses

Both evaporation and infiltration are simulated on the porous pavement. Evaporation can be entered as a constant value, as a time series, as a climate file, or as monthly average values.

For this template, average daily evaporation rates for each month were obtained for Corvallis, Oregon from the U.S. Bureau of Reclamation's AgriMet website. These values are shown in Figure A-5. Changes to the evaporation values are performed in the same manner discussed in Section A.2.1.4.

Infiltration occurring on the porous pavement is simulated in SWMM by using the Horton infiltration model. The maximum and minimum infiltration rates are 3 inches per hour and 0.5 inch per hour, respectively. Changes to these parameters are made by opening the properties window for the infiltration trench subcatchment, highlighting the Infiltration field, and clicking on the ellipsis button.

APPENDIX B

MODEL GRAPHICAL USER INTERFACES

B.1 SWMM5 GUI

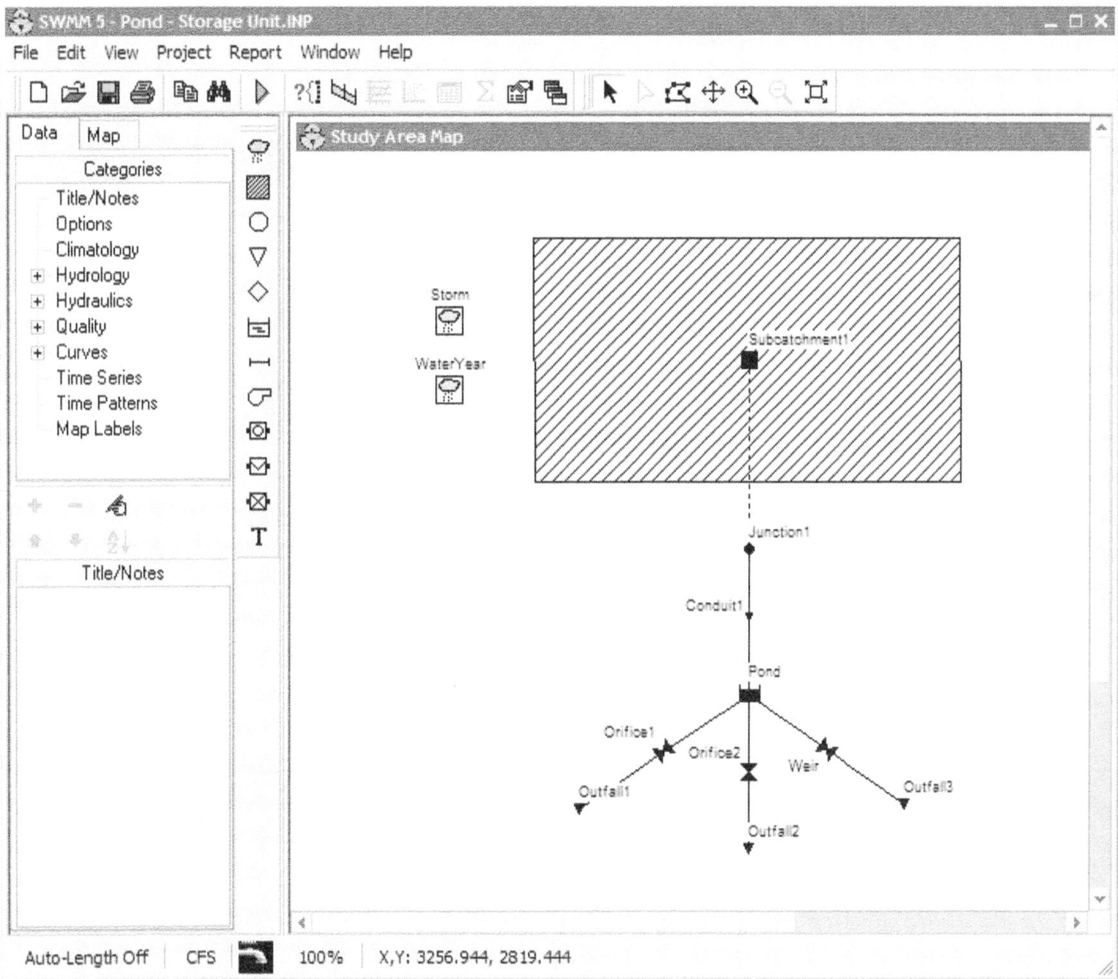

B.2 WinTR-55 GUI

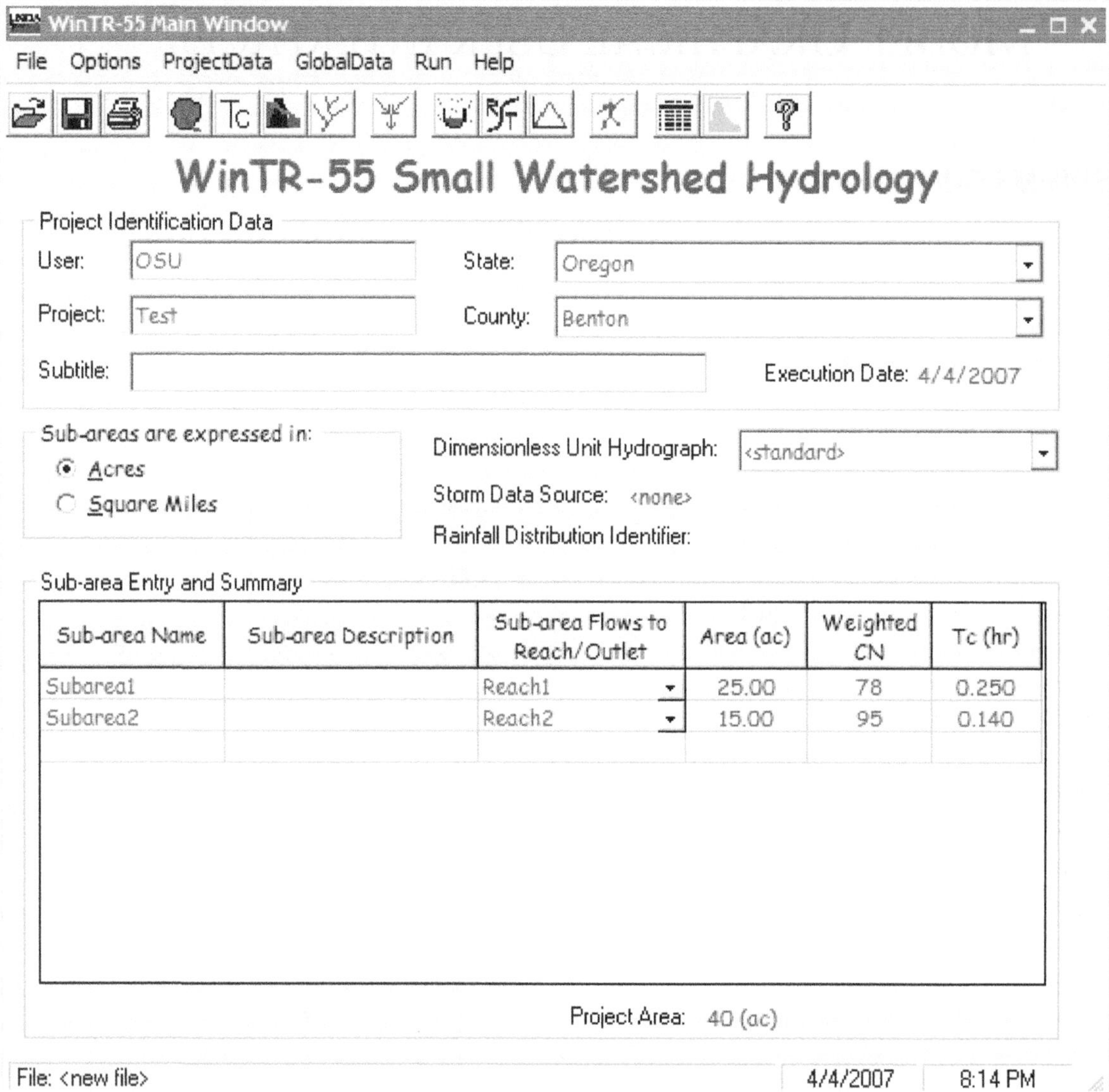

B.3 WinHSPF GUI

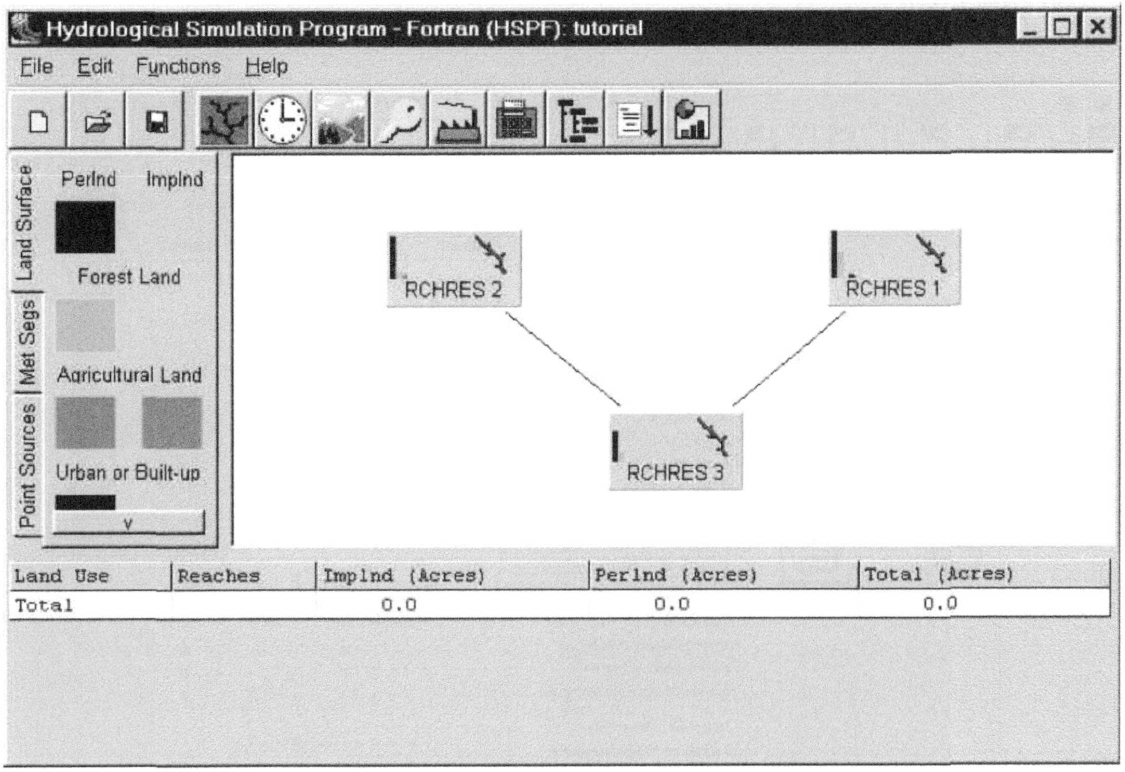

B.4 WWHM3 GUI

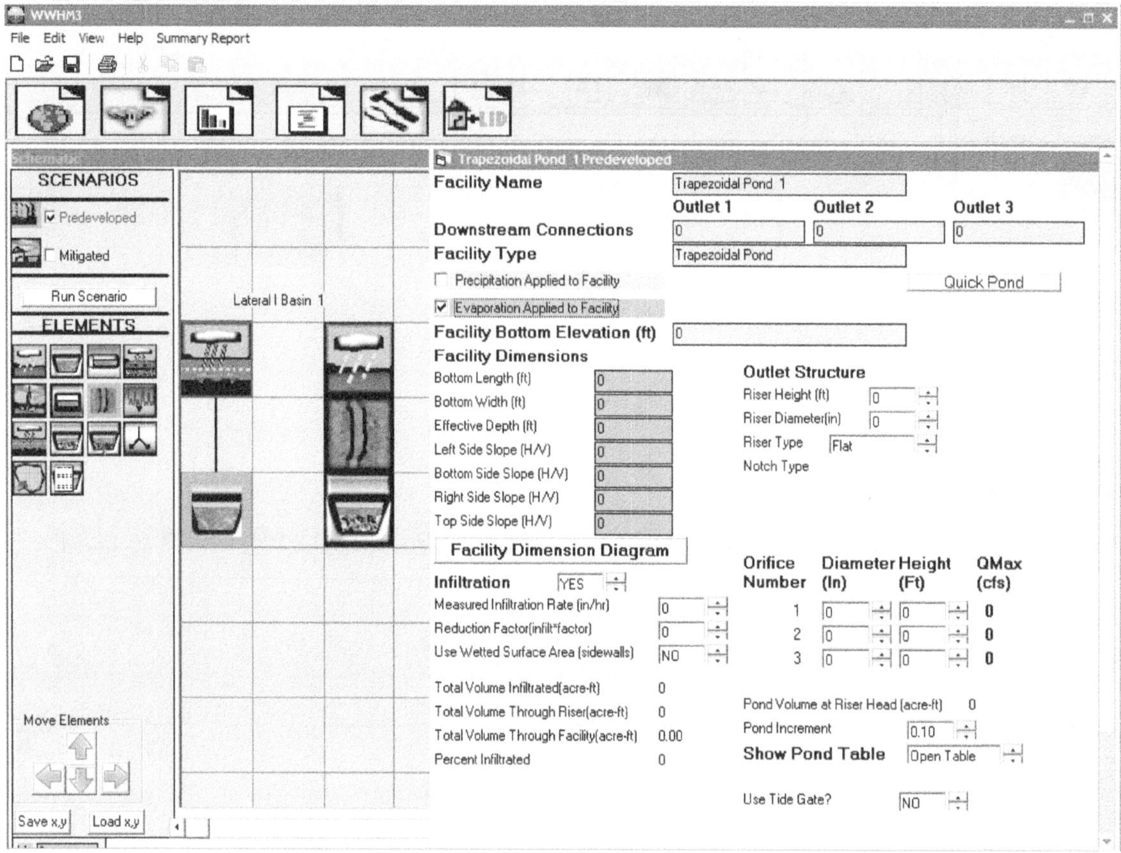
WWHM3
File Edit View Help Summary Report
Schematic
SCENARIOS
Predeveloped
Mitigated
Run Scenario
ELEMENTS
Move Elements
Save x,y
Load x,y
Lateral I Basin 1
Trapezoidal Pond 1 Predeveloped
Facility Name
Trapezoidal Pond 1
Outlet 1
Outlet 2
Outlet 3
Downstream Connections
Facility Type
Trapezoidal Pond
Precipitation Applied to Facility
Quick Pond
Evaporation Applied to Facility
Facility Bottom Elevation (ft)
Facility Dimensions
Bottom Length (ft)
Bottom Width (ft)
Effective Depth (ft)
Left Side Slope (H/V)
Bottom Side Slope (H/V)
Right Side Slope (H/V)
Top Side Slope (H/V)
Outlet Structure
Riser Height (ft)
Riser Diameter(in)
Riser Type
Flat
Notch Type
Facility Dimension Diagram
Infiltration
YES
Measured Infiltration Rate (in/hr)
Reduction Factor(infilt*factor)
Use Wetted Surface Area (sidewalls)
NO
Orifice Number
Diameter (In)
Height (Ft)
QMax (cfs)
Total Volume Infiltrated(acre-ft)
Total Volume Through Riser(acre-ft)
Total Volume Through Facility(acre-ft)
0.00
Percent Infiltrated
Pond Volume at Riser Head (acre-ft)
Pond Increment
0.10
Show Pond Table
Open Table
Use Tide Gate?
NO

APPENDIX C

BIORETENTION MONITORING IN WASHINGTON, D.C.

C.1 Bioretention Monitoring

Modeling, as discussed previously in this report, is a critical planning component for the implementation of decentralized controls. Supporting modeling with quantifiable performance data optimizes model predictions and highlights the advantages of decentralized systems. Two bioretention systems in Washington, D.C., were modeled to assess their performance as urban BMPs. Although design of one system was not optimal, monitoring of both indicates that this type of decentralized control is effective in urban retrofit applications. Because of the interest in applying these systems to combined sewer areas and to reduce loadings from MS4 systems, volume management is a predominant concern; water quality monitoring was undertaken in this effort to determine both the effectiveness of bioretention at removing stormwater pollutants and as a means of comparing performance with other conventional urban BMPs. The presented results are from field sampling 15 events at the bioretention site at the Benning Road Bridge over the Anacostia River and a field study of a bioretention system located at a paved parking lot at the Washington Navy Yard. Figures C-1 and C-2 show the monitoring sites. Influent and effluent samples were collected; only major rainfall events, greater than 0.1 inch with an antecedent dry period of 72 hours, in compliance with U.S. EPA regulations for further analysis in the laboratory, were monitored.

This study monitored the efficiency of bioretention in terms of pollutant removal. The results section presents the data in tabular form. All values presented are the average of three measurements, with the standard deviation for that average presented in the adjacent column. In general, bioretention is efficient in terms of pollutant removal in the following order.

Table C-1. Summary of Bioretention Monitoring Results.

Parameter	Removal Efficiency (%)
Total suspended solids	98
Zn	80
Cu	75
Pb	71
Cd	70
NH_3-N	65
Fe	51
Cr	42
NO_2^--N	27
Al	17
PO_4^{3-}-P	3

Figure C-1. Benning Road Bioretention Site.

Figure C-2. Washington Navy Yard Bioretention.

Water quality enhancement occurs through the bioretention facility via biological, chemical, and physical processes including phytoremediation, precipitation, adsorption, complexation, microbial activity, decomposition, sedimentation, filtration, and volatilization. In urban environments these systems are typically designed by filling a concrete box with gravel, sand, planting soil, a top layer of mulch, and various species of water loving plants.

To capture the highest concentration of water pollutants the focus is normally on "first flush" discharge samples thus, samples are normally collected after a major storm event after at least 72 hours of extended dry period has elapsed from one event to another. Rain events are monitored by checking the local weather station or weather-related internet sites. Stormwater runoff samples from the all sites are collected into cleaned 4-liter jars for further analysis in the lab.

BMPs are part of U.S. EPA's effort to regulate the release of dangerous chemicals into natural aquatic environments through water quality standards set forth by the National Pollutant Discharge Elimination System (NPDES). In Washington, D.C., compliance to the NPDES standards is determined by the level of compliance to the limits specified by the District's Water and Sewer Regulation. To meet the permitted pollutant levels, BMPs such as sand filters, bioretention basins, dry and wet ponds, and constructed wetlands are installed to treat stormwater runoff from parking lots, highways, rooftops, lawns, construction sites, and industrial facilities. Pollutants targeted for removal include nutrients (nitrogen, phosphorus, and biochemical oxygen demand) from domestic wastewater, suspended solids from streets and construction sites, and debris from streets.

The District of Columbia requires pollutant removal in stormwater in compliance with U.S. EPA's recommendations for discharge of compounds listed as priority pollutants. U.S. EPA regulations require elimination of toxic heavy metals and toxic organic compounds from wastewater and stormwater entering natural streams. Additionally, water entering the natural streams must not raise or lower the pH outside the natural range of 6.0-9.0 and must not alter the temperature beyond or below the 60-140° F range.

C.1.1 Sources of Pollution

In this report, pollution is mostly associated with certain types of activities and land use. Sources of pollution in the Washington, D.C., and surrounding areas include:

- Leaking gasoline tanks that release toxic petrochemical compounds into the environment;
- Surface runoff from parking lots and highways that contain oils, anti-freeze, and petroleum products from automobiles;
- Constructions sites that generate high sediment payload in stormwater;
- Road maintenance;
- Snow removal operations that generate highly corrosive and toxic chemicals from pre-wetting and anti-icing agents; and
- Infrequent and unprotected industrial chemicals storage sites.

The highest proportion of stormwater pollution within the city limits emanates from impermeable concrete surfaces such as parking lots, highways, and bridges. A significant amount of pollution originates from rooftops and lawns, but the pollutant concentrations in these areas decrease rapidly with increased rainfall frequency within a season.

C.1.2 Status of BMPs in Washington, D.C.

The majority of the transportation-related BMPs in the Washington, D.C., are water quality sand filters. The systems are not monitored regularly resulting in a large number either being forgotten or not receiving maintenance for a long time. These systems become a pollution concern especially because they are not self-purifying. Currently, there has been renewed interest in increasing the number of decentralized practices such as tree boxes and bioretention basins, which offer a certain degree of self-purification and pollutant bioremediation. The use of wetlands so far has been mostly in stream restoration projects.

C.1.3 Recommendations

The recommendations presented here are based on the currently available information on the BMPs installed in the greater Washington, D.C, area. These recommendations are drawn from a relatively small representative sample of BMPs for which operational data was available:

C.1.3.1 The Use of Biological Systems

Most of the BMPs currently installed in Washington, D.C., use physicochemical operations that include straining action of filtration devices, surface adsorption, precipitation, and sedimentation processes to remove pollutants in stormwater. These systems eventually reach catalytic saturation after removing a determinate amount of pollutants such that backwashing or other forms of maintenance are required at regular intervals to re-establish system integrity. Biological systems, on the other hand, are technically self-cleansing in that pollutants are either degraded into less toxic forms or are taken up by plants and microorganisms growing in the system.

Currently, apart from a few experimental units situated at the Washington Navy Yard and at Wheeler Creek, 900 Varney Street SE, there is insignificant biological system usage in the area. Wetland systems are predominantly used in stream bank and river ecosystem restoration projects. The absence of biological systems in the stormwater management practices is probably because of lack of previous experience with the systems, the systems being relatively new and in many cases experimental.

C.1.3.2 Monitoring and Reporting

In this study, recommendations are drawn for the continuous monitoring and regular reporting of installed BMPs for the following reasons:

- The need to create and maintain local and national BMP databases accessible by research organizations and the public. One such database is the National Stormwater Best Management Practices Database implemented and managed by the American Society of Civil Engineers and U.S. EPA. Presently, data in the National Database related to the BMPs and BMP performance in the Washington, D.C., area is sparse.
- To coordinate stormwater management and watershed conservation efforts by different government agencies and other organizations working to improve the water

quality of the Rock Creek, Anacostia and Potomac Rivers, and the Chesapeake Bay through effective stormwater pollution control.

- To update the total maximum daily loads used for water quality modeling and for formulating criteria for pollutant discharge permitting in Greater Washington, D.C.
- The current role that the Department of Public Works, Division of Transportation, plays in the management of BMPs at its facilities is deemed acceptable when compared with other large urban municipalities throughout the United States. However, as mentioned above, very little assessment of the quality of pollutant removal efficiency and the resulting improvement on water quality has been performed.

Presently, evaluation of performance of all the BMPs in the area is difficult because of scarcity of reported data.

C.1.4 Results

There are many variables that cause fluctuations in the performance of stormwater devices. When monitoring the devices from varying sites there are no set rules that reveal what determines good versus bad performance. Unlike in drinking water and wastewater treatment, the guidelines for stormwater devices do not call for a given concentration to be met, but rather typically are designed to retain a certain amount of flow.

An evaluation of BMP performance can be derived from a comparison between the values found in the U.S. EPA's priority and non-priority pollutants for fresh water, and the values found going into and out of the monitored sites is more appropriate. With these values, that take into account the impact the pollutants have on aquatic life and the possibility of the use of the water as drinking water downstream, the heavy metal concentrations coming from the sites are of larger concern. Many of the parameters found on the priority pollutants list are found in excess going into and out of the sites monitored.

Table C-2. Chronic Priority Pollutants for Fresh Water (U.S. EPA).

Pollutant	Chronic Concentration (ppb)
Copper	9
Cadmium	0.25
Zinc	120
Chromium	11
Lead	2.5
Arsenic	150
Mercury	0.77

Table C-3. Non-Priority Pollutants (U.S. EPA).

Pollutant	Concentration (mg/L)
pH	6.5-9.0
Dissolved oxygen	Dependent
Temperature	Dependent
Total suspended solids	80
Total dissolved solids	250
Polycyclic aromatic hydrocarbons	0
Total nitrogen	10
Total phosphorus	0.1

Comparing the values found in Tables C-3 and C-4 with the values for the measured parameters in Table C-5, which contains the summary of results for the Benning Road bioretention, it is evident that several parameters are excessive. Copper, cadmium, lead, and mercury had average concentrations into and out of the BMP in excess of the priority pollutant concentrations. All of the non-priority pollutants were in a reasonable range of values, however the efficiency of removal did not compare well to other bioretention studies.

Table C-4. Summary of Bioretention Data.

Contaminant	Influent Average and Standard Deviation.	Effluent Average and Standard Deviation	Removal Efficiency
pH	8.09 ± 0.46	7.75 ± 0.46	
Dissolved oxygen	5.87 ± 1.87	5.80 ± 0.88	
Temperature (°C)	19 ± 3.5	18.3 ± 4.1	
Total suspended solids	176 ± 396	24 ± 28	*86%*
Total dissolved solids	146 ± 353	14 ± 24	*91%*
Cu	23 ± 29	11 ± 29	*53%*
Cd	9 ± 32	2 ± 9	*78%*
Zn	70 ± 32	52 ± 27	*26%*
Cr	10 ± 7	5 ± 4	*50%*
Pb	47 ± 179	16 ± 56	*66%*
As	29 ± 112	31 ± 119	*-6%*
Hg	54 ± 126	42 ± 86	*22%*
TP	0.5 ± 0.4	0.2 ± 0.2	*60%*
PO_4^{-3}	0.0 ± 0.1	0.0 ± 0.1	*0%*
NO_2^-	0.4 ± 0.6	0.3 ± 0.5	*-25%*
NO_3^-	0.2 ± 0.3	0.2 ± 0.2	*0%*
NH_4^+	22 ± 55	16 ± 40	*27%*
COD	*112 ± 92*	*66 ± 43*	*41%*

The Navy Yard bioretention field study was conducted to evaluate the efficiency of the system in terms of dissolved oxygen, temperature, pH, and heavy metals removal. In Figure C-3 the dissolved oxygen concentration of bioretention influent and effluent are presented. Given the low temperature of rain events 4 through 6, which marked the end of winter, the dissolved oxygen concentration was supersaturated leading to high dissolved oxygen concentrations. Even at supersaturated concentrations, the dissolved oxygen concentration was higher as the stormwater left the bioretention cell for all of the storm events. The pH of rain water, influent, and effluent stormwater are presented in Figure C-4. The slightly acidic nature of rainfall in the Washington, D.C., area can be seen with an average pH of 5.4 ± 0.2. The area of this study is known to have acid rain with a pH as low as 4.2-4.4. Nevertheless the pH of the rain water increased as it created runoff into the system with an average pH of 7.5 ± 0.5 and did not significantly change as the stormwater passed through the system resulting in an effluent pH of 7.4 ± 0.4.

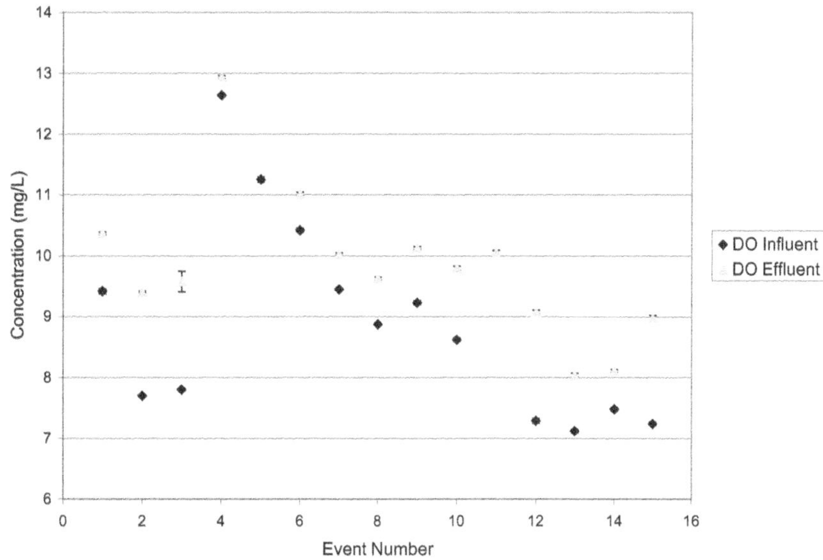

Figure C-3. Navy Yard Bioretention—Dissolved Oxygen Concentration for 15 Storm Events.

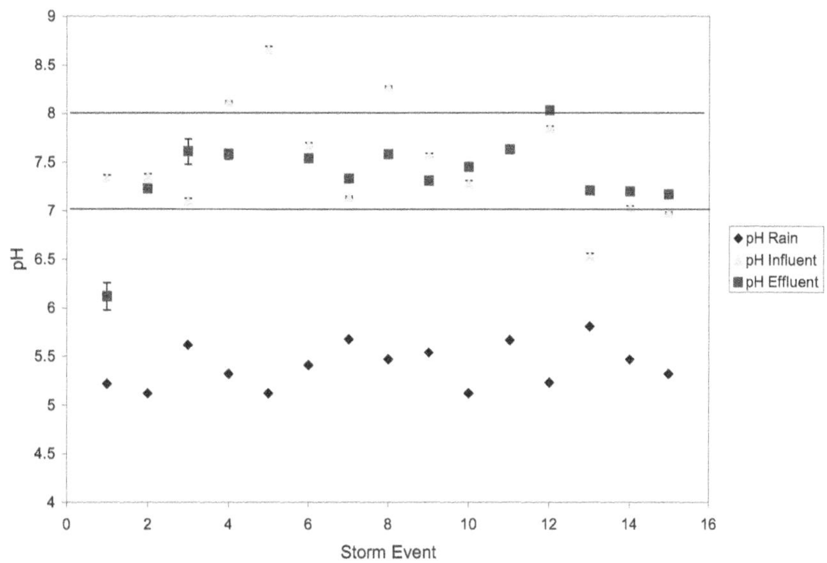

Figure C-4. Navy Yard Bioretention—pH of Rain Water Influent and Effluent.

The summary of results for the average concentrations of heavy metals in the influent and effluent are presented in the Table C-5. In addition, the drinking water maximum contaminant level required by U.S. EPA is listed as a reference. In this study only lead and cadmium concentrations in the influent exceeded the drinking water standard. Nevertheless, the effluent metal concentrations were observed to be below the influent concentrations for all of the heavy metals displaying the successful adsorption of these compounds in the bioretention cell.

Table C-5. Navy Yard Bioretention—Summary of Heavy Metal Concentrations.

	Cu	Cd	Zn	Cr	Pb	Al	As	Fe
C_{in} (µg/L)	56.8	5.6	98.3	7.6	41.4	3.5	2	26.8
C_{ef} (µg/L)	10.9	1.9	20.6	3.6	10.2	2.9	1.7	12.6
Removal %	81	66	79	53	75	17	11	53
Maximum contaminant level (g/L	1000	5	500	100	15	200	50	300
Detection limit (µg/L)	0.85	0.065	0.05	0.35	1.5	1.55	2	0.6

In Figure C-5, the removal of cadmium is depicted for the 15 storm events evaluated in this investigation. The bioretention performed well, removing cadmium to below the drinking water maximum contaminant level in all but two storm events, which happened to correspond to the events with the highest concentrations of cadmium. The percentage removal of cadmium based on concentration was 66% and 62% based on the load to the system. Figure C-5 shows the drastic variation in the influent concentration in to the bioretention from storm to storm. This is largely dependent on the varying amounts of deposition in between storm events and was true for all of the heavy metals monitored in this study.

Lead was the only other heavy metal consistently present at influent concentrations that exceeded the maximum contaminant level for drinking water. All of the storm events contained concentrations of lead in excess of the drinking water standard. The bioretention cell removed lead to concentrations below the drinking water standard for all storm events with an average removal rate of 75%. Copper, zinc, iron, chromium, aluminium, and arsenic were removed from the stormwater at average percentage removal rates of 81%, 79%, 53%, 53%, 17%, and 11%, respectively.

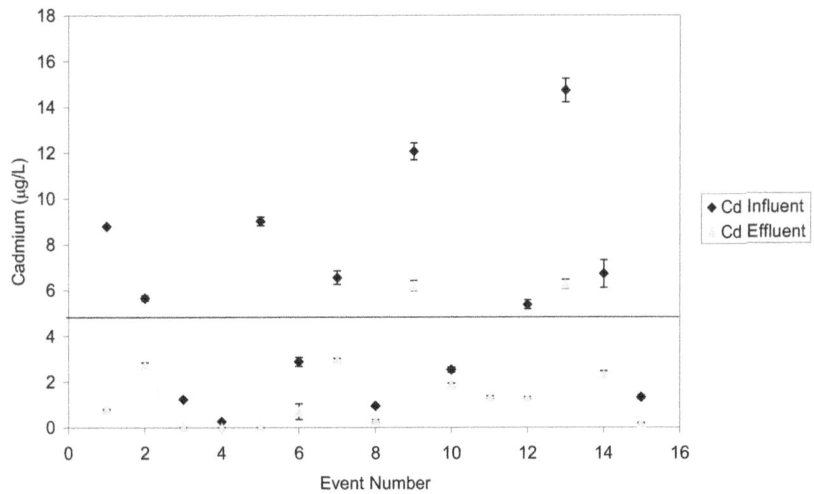

Figure C-5. Navy Yard Bioretention—Cadmium Concentration for 15 Storm Events.

Nutrient removal in the field study showed successful removal of nitrate, nitrite, and ammonia, however none of these parameters were ever present at concentrations exceeding 1 mg/L. Soluble phosphorus as phosphate was never detected in the influent or effluent from the bioretention cell, with a detection limit of 0.1 mg/L. Ammonia was present in the influent at an average concentration of 0.5 mg/L and was never detected in the effluent. Nitrate was not detected in 8 of the 15 storm events and was only detected in the effluent of four events. Nutrients were not present at elevated concentrations in this parking lot.

The metal removal efficiency results were less than those found in the literature for laboratory studies by Davis (Davis, 2001; Davis et al., 2003). This is certainly because of the lack of maintenance of the bioretention mulch layer, which should retain metals at efficiencies from 88-99%. Laboratory and pilot studies have shown that the surface mulch layer is the most important component of the bioretention system for metal removal. In this study no special changes were made from the normal maintenance and operation of the bioretention area studied. During the early spring/late winter timeframe of this study the bioretention area contained a low growth of plants and had not received new mulch. The facility was two years old and had not been replanted in the time prior to the study. The plant roots should have improved metal removal and the bioretention age should have provided more efficient metal uptake. However, in many of the earlier studies a constantly higher concentration of synthetic water was used instead of the changing concentrations of actual stormwater.

C.1.5 Conclusions

Benning Road Bioretention

In this study, grab samples of the influent and effluent water for 15 storm events have been monitored for the bioretention site located adjacent to the Benning Road Bridge. The monitoring results and site inspection suggest that the Benning Road bioretention site was poorly designed, constructed, maintained, or all of the above. From visual inspection of the catch basins, one of the inlets to the bioretention did not operate properly for the length of the project. Only a trickle of storm water reaches the bioretention site from the clogged inlet, the majority of the storm water proceeds directly into the combined sewer system with no diversion to the bioretention cell from one of the two catch basins. This is most likely a maintenance problem.

In addition, this site seems to be too small for the watershed that it is theoretically supposed to be treating. For 7 of the 15 storm events monitored the overflow of the bioretention was reached. This bioretention system seems to overflow for relatively small storms and it is only treating the inflow of one of the two pipes that are connected to it. Pollutants that are dissolved pass through the system with little potential for removal when overflow occurs.

In summary, when evaluating the average values of the parameters in the inlet and outlet from this bioretention, it must be concluded that it is not representative of the capability of this technology when properly designed, built, and maintained. Only total suspended solids and total dissolved solids were removed on a semi-consistent basis from the system, at 86% and 91% removal efficiencies. Most of the dissolved constituents were removed at no better than 50% efficiency. These values are much less than removal percentages of bioretention systems in the literature or the performance of other bioretention systems monitored in the past.

Navy Yard Bioretention

The bioretention removed pollutants with the following efficiencies: TSS (~ 98%) > Zn (~ 80%) > Cu (~ 75%) > Pb (~ 71%) > Cd (~ 70%) > NH_3-N (~65%) > Fe (~ 51%) > Cr (~ 42%) > NO^-_2-N (~ 27%) > Al (~ 17%) > PO^{3-}_4-P (~3%). From the field results Cu (II), Zn (II), and Pb (II) were removed significantly at 81 ± 29%, 79 ± 14% and 75 ± 19%. The field results indicate that bioretention facilities can be effective for the removal of heavy metals, nutrients, and increase the dissolved oxygen content. To ensure removal of heavy metals to 90%, the soil organic matter must be kept high and plants must be well maintained.

References

American Forests. *Urban Ecosystem Analysis Delaware Valley Region; Calculating the Value of Nature*; American Forests: Washington, D.C., 2003.

American Society of Civil Engineers. *Report Card for America's Infrastructure,* ASCE: Reston, VA, 2005, pp. 4–12. Available at: http://www.asce.org/reportcard/2005/index.cfm.

American Society of Landscape Architects. *Green Roof Demonstration Project Fact Sheet*; ASLA: Washington, D.C., 2006.

Anacostia Waterfront Corporation. *Environmental Standards*; Anacostia Waterfront Corporation Environmental Standards Development Committee: Washington, D.C., 2007.

Banting, D.; Doshi, H.; Li, J.; Missios, P.; Au Beth, A.; Currie, B.A; Verrati, M. *Report on the Environmental Benefits and Costs of Green Roof Technology for the City of Toronto.* City of Toronto and Ontario Centres of Excellence—Earth and Environmental Technologies: Toronto, ON, Canada, 2005.

Barrett, M.E.; Malina, Jr., J.F.; Charbeneau, R.J.; Ward, G.H. Characterization of Highway Runoff in Austin, Texas. *J. Environ. Engin.* **1998,** *124*(2), 131–137.

Bay Area Stormwater Management Agencies Association. *Using Site Design Techniques to Meet Development Standards for Stormwater Quality: A Companion Document to Start at the Source*; BASMAA: San Francisco, 2003.

Beach, D. *Coastal Sprawl: The Effects of Urban Design on Aquatic Ecosystems in the United States*; Pew Ocean Commission: Washington, D.C., 2002.

Becker, M.; Raasch, U. Sustainable Rainwater Management in the Emscher River Catchment Area. *Water Sci. Technol.* **2002**, *45*(3). (Selected Proceedings of the 2nd International Conference on Interactions Between Sewers, Treatment Plants, and Receiving Waters In Urban Areas, Held In Lisbon, Portugal, 19-22 February 2001),

Bedient, P.B.; Huber, W.C.; Vieux, B.E. *Hydrology and Floodplain Analysis*, 4th ed.; Prentice Hall: Upper Saddle River, NJ, 2008.

Beyerlein, D. Flow Duration-Based Stormwater Mitigation Modeling. *Stormwater* **2005**, *6*(3). Available at: http://www.stormh2o.com/may-june-2005/flow-mitigation-modeling.aspx

Bicknell, B.R.; Imhoff, J.C.; Kittle Jr., J.L.; Donigian, Jr., A.S.; Johanson, R.C *Hydrological Simulation Program - FORTRAN User's Manual for Release 10*; U.S. EPA: Athens, GA, 1993.

Bicknell, B.R.; Imhoff, J.C.; Kittle Jr., J.L.; Donigian Jr., A.S.; Johanson, R.C. *Hydrological Simulation Program – FORTRAN; HSPF Version 12 User's Manual*. U.S. EPA: Athens, GA, 2001.

Brown, S. *Pennsylvania Standards for Residential Site Development*, Project No. 04-DEC-12SG; Pennsylvania Housing Research Center, Pennsylvania State University: University Park, PA, 2007. Available at: http://www.engr.psu.edu/phrc/Land%20Development%20Standards.htm (accessed January 2008).

Brown, S.; Stein, S.; Saccone, D.; Butler, L.; Williamson, S.; Weinkam C. *Watershed Management Plan for Reston, VA*; GKY and Associates, Inc.: Springfield, VA, 2002.

Casey Trees and LimnoTech. *The Green-Build-out Model: Quantifying the Stormwater Management Benefits of Trees and Green Roofs in Washington, D.C.*; Casey Trees and LimnoTech: Washington, D.C., 2007.

Chesapeake Bay Watershed Blue Ribbon Finance Panel. *Saving a National Treasure: Financing the Cleanup of the Chesapeake Bay—A Report to the Chesapeake Executive Council*; Chesapeake Bay Program: Annapolis, MD, 2004.

Chesapeake Executive Council. *Directive No. 01-1 Managing Storm Water on State, Federal and District-Owned Lands and Facilities*; Chesapeake Bay Program: Annapolis, MD, 2001. Available at: http://www.chesapeakebay.net/pubs/stormwater_directive_120301.pdf (accessed August 2007).

City of Chicago Department of Environment. *Urban Heat Island Initiative Alley Reconstruction*; Department of Environment: Chicago, 2005. Available at: http://egov.cityofchicago.org/webportal/COCWebPortal/COC_ATTACH/UHIAlley.pdf (accessed January 2005).

Clean Water Services. *Design and Construction Standards for Sanitary Sewer and Surface Water Management;* Clean Water Services: Hillsboro, OR, 2004.

Clear Creek Solutions, Inc. *Western Washington Hydrology Model, Version 3.0, User Manual;* Clear Creek Solutions, Inc.: Mill Creek, WA, 2006. Available at: http://www.clearcreeksolutions.info/ftp/public/downloads/WWHM3_User_Manual.pdf.

Currie, B.A.; Bass, B. *Estimates of Air Pollution Mitigation with Green Plants and Green Roofs Using the UFORE Model;* Environment Canada, Adaptation and Impacts Research Group, Gatineau, PQ, Canada, 2005.

Davis, A.P.; Shokouhian, M.; Sharma, H.; Minami, C. Laboratory Study of Bioretention for Urban Stormwater Management. *Water Envir. Res.*, **2001** *73*.

Davis, A.P.; Shokouhian, M.; Sharma, H.; Minami, C.; Winogradoff, D. Water Quality Improvement Through Bioretention: Lead, Copper, and Zinc Removal. *Water Envir. Res*. **2003,** *75*, 1.

Derry, B.; Butchart, C.; Graham, P. *Analysis and Recommendation for the use of LID Techniques in Puget Sound*; CH2M Hill Technical Memorandum No. 2; Puget Sound Action Team: Olympia, WA, 2004. Available at http://www.psparchives.com/publications/our_work/stormwater/lid/development_tech_materials/PSAT_TechMemo2.pdf.

District of Columbia Water and Sewer Authority. *Combined Sewer System Long Term Control Plan*; WASA: Washington, D.C., 2002.

Donigian, A.S.; Crawford, N.H. *Modeling Nonpoint Pollution from the Land Surface,* EPA-600/3-76-083; U.S. EPA, Office of Research and Development: Cincinnati, OH, 1976.

Duluth Streams. *Stormwater Management Plan*. Available at http://www.duluthstreams.org/stormwater/duluth/stormwater_plan.html.

Ecosystem Valuation Website. http://www.ecosystemvaluation.org/big_picture.htm (accessed February 2007).

Environmental Valuation and Cost Benefit Website. http://www.costbenefitanalysis.org/glossary.htm (accessed November 2006).

Federal Register **1994**, *59*, 18688.

Fitzgerald, S. Berlin: A Decentralized Approach to Sustainability, 2006. Available at: http://courses.umass.edu/greenurb/2006/sfitzgerald/Innovations_GAF.htm (accessed December 2008).

Grant, G., Engleback, L., and Nicholson, B. (2003). Green Roofs: Their Existing Status and Potential for Conserving Biodiversity in Urban Areas, Report Number 498. English Nature Research Reports.

Guntner, A.; Olsson, J.; Calver, A.; Gannon, B. Cascade-Based Disaggregation of Continuous Rainfall Time Series. *Hydrol. Earth Syst. Sci.* **2001** *5*(2), 145–164. Available at: http://www.copernicus.org/EGU/hess/hess.html (accessed August 2007).

Horner, R.R.; Lim, H.; Burges, S.J. *Hydrologic Monitoring of the Seattle Ultra-Urban Stormwater Management Projects: Summary of the 2000-2003 Water Years;* Water Resources Series: Technical Report No. 181; University of Washington Department of Civil and Environmental Engineering: Seattle, WA, 2004.

Huber, W.C.; Cannon, L.; Stouder, M. *BMP Modeling Concepts and Simulation,* EPA/600/R-06/033, U.S. EPA: Cincinnati, OH, 2006. Available at: http://www.epa.gov/ORD/NRMRL/pubs/600r06033/epa600r06033toc.pdf.

Huber, W.C.; Dickinson, R.E. *Storm Water Management Model Version 4, User's Manual,* EPA/600/3-88/001a (NTIS PB88-236641/AS); U.W. EPA: Athens, GA, 1968.

Kaspersen, J. The Stormwater Utility: Will it Work in Your Community? *Stormwater J. Surf. Water Qual. Profess.* **2000,** *2.*

Lambert, A. *Economic Valuation of Wetlands: an Important Component of Wetland Management Strategies at the River Basin Scale.* Ramsar Convention on Wetlands, Rue Mauverney 28, CH-1196 Gland, Switzerland, 2003. Available at: http://www.ramsar.org/features/features_econ_val1.htm

MacMullen, E. *Using Benefit-Cost Analyses to Assess Low-Impact Developments.* Presentation abstract for the Second National Low Impact Development Conference, Wilmington, NC, March 12–14, 2007.

Massachusetts Office of Coastal Zone Management. *The Coastal Nonpoint Source Pollution Control Program in Massachusetts—The Planning Process: Bringing the Right Groups Together;* Office of Coastal Zone Management: Boston, MA, 1994. Available at: www.mass.gov/czm/mpsplpro.htm (accessed August 2007).

Miller, C. *Green Roof Benefits.* Roofscapes, Inc.: Philadelphia, PA, 2004. Available at: http://www.roofmeadows.com/technical/benefits.shtml (accessed August 2007).

National Association of Flood and Stormwater Management Agencies. *Guidance for Stormwater Funding;* NAFSMA: Washington, D.C., 2006.

National Research Council of the National Academies. *Valuing Ecosystem Services: Toward Better Environmental Decision-Making*; National Academy Press: Washington, D.C., 2004.

Natural Resources Conservation Service. *Urban Hydrology for Small Watersheds: TR-55;* U.S. Department of Agriculture, NRCS: Washington, D.C., 1986.

Natural Resources Conservation Service. *WinTR-55 User Manual*; U.S. Department of Agriculture, NRCS: Washington, D.C., 2002. Available at: http://www.wsi.nrcs.usda.gov/products/W2Q/H&H/Tools_Models/WinTR55.html

Natural Resources Defense Council. *Out of the Gutter: Reducing Polluted Runoff in the District of Columbia;* NRDC: New York, NY, 2002.

Natural Resources Defense Council. *Rooftops to Rivers: Green Strategies for Controlling Stormwater and Combined Sewer Overflows*; NRDC: New York, NY, 2006.

Newman, P.; Kenworthy, J. *Sustainability and Cities: Overcoming Automobile Dependence.* Island Press: Washington, D.C., 1999.

Novotny, V.; Brown, P.R. *Cities of the Future: Towards Integrated Sustainable Water and Landscape Management*; IWA Publishing: London, UK, 2007.

Perello, J.; DePalma, S. Park and Ramble: Sayen Gardens Transformed. *Landsc Arch Specifier News*, **2006**.

Philips, J.R. (1957). The Theory of Infiltration: The Infiltration Equation and Its Solution, Soil *Science* **1957**, *53*, 345–375.

Powell, L.M.; Rohr, E.S.; Canes, M.E.; Cornet, J.L.; Dzuray, E.J.; McDougle, L.M. *Low-Impact Development Strategies and Tools for Local Governments: Building a Business Case*, Report Lid50t1; LMI Government Consulting: McClean, VA, 2005.

Rawls, W.J.; Brakensiek, D.L.; Miller, N. Green-Ampt Infiltration Parameters from Soils Data, J. Hydraulic Engineering, *ASCE* **1983**, *109*(1), 62–70.

Richman, T.; Dawe, P.; Worth, J.; Aldrich, J.; Quasebarth, T.; Endicott, J.; Ferguson, B.K. *Start at the Source: Design Guidance Manual for Stormwater Quality Protection;* Bay Area Stormwater Management Agencies Association: San Francisco, 1999.

Rossman, L.A. *Storm Water Management Model User's Manual Version 5.0;* U.S. EPA: Cincinnati, OH, 2005. Available at: http://www.epa.gov/ednnrmrl/models/swmm/index.htm.

Salzman, J.; Thompson Jr., B.H. *Environmental Law and Policy;* Foundation Press. New York, NY, 2003.

Seattle Public Utilities. *Triple Bottom Line Guidebook: A Reference Guide for PDP Analyses*; Seattle Public Utilities: Seattle, WA, 2005.

Steed, A. *Naturalized Streetscapes: A Case Study of Crown Street, Vancouver*. Greenways – City of Vancouver. Available at: http://www.sustainabilitynow.com/Docs/Naturalized%20Streetscapes-AS.pdf?CFID=9810761&CFTOKEN=66307383. (accessed January 2009).

University of Maryland Environmental Finance Center. *Schuylkill Action Network Financing Strategy;* University of Maryland: College Park, MD, 2007.

University of North Carolina Environmental Finance Center Website. http://www.efc.unc.edu/projects/arcregional.htm (accessed March 2007).

U.S. Environmental Protection Agency. *Clean Watersheds Needs Survey 2004 Report to Congress*; U.S. EPA, Office of Wastewater Management: Washington, D.C., 2008.

U.S. Environmental Protection Agency. *Combined Sewer Overflows: Guidance for Long-Term Control Plan;* EPA 832-B-95-002; U.S. EPA, Office of Water: Washington, D.C., 1995.

U.S. Environmental Protection Agency. Many Paths Lead to the Adoption of Low Impact Development. *Nonpoint Source News-Notes*. **2006**, *76*.

U.S. Environmental Protection Agency. *Reducing Stormwater Costs through Low Impact Development (LID) Strategies and Practices,* EPA 841-F-07-006; U.S. EPA: Washington, D.C., 2007.

U.S. Geological Survey. *Water Quality and Nonpoint Sources in Urban Watersheds: Key Findings from the First Decade of NAWQA Studies;* National Water-Quality Assessment (NAWQA) Program, US Geological Survey: Reston, VA, 2008. Available at: http://water.usgs.gov/nawqa/informing/urbanization.html (accessed January 2008).
U.S. Green Building Council. *LEED-NC*: *Green Building Rating System for New Construction & Major Renovations Version 2.2*; USGBC: Washington, D.C., 2005. Available at: http://www.usgbc.org (accessed September 2007).

Vingarzan, R.; Taylor, B. *Trend Analysis of Ground Level Ozone in the Greater Vancouver / Fraser Valley Area of British Columbia*; Environment Canada—Aquatic and Atmospheric Sciences Division: Gatineau, PQ, Canada, 2003.

Wachter, S. M.; Gillen, K.C. *Public Investment Strategies: How They Matter for Neighborhoods in Philadelphia, Philadelphia, PA*: Wharton School of Business, University of Pennsylvania: Philadelphia, PA, 2006. Available at http://www.community-wealth.org/_pdfs/articles-publications/state-local/paper-wachter-gillen.pdf.

Wilma, D. Seattle City Light Inaugurates Kill-A-Watt to Conserve Electricity on July 17, 1973; *HistoryLink: the Online Encyclopedia of Washington State History*, 2000. Available at: http://www.historylink.org/essays/output.cfm?file_id=3647 (accessed October 2007).

Wilma, D. Seattle City Council Rejects Nuclear Power in Favor of Conservation on July 12, 1976; *HistoryLink: the Online Encyclopedia of Washington State History*, 2001. Available at http://www.historylink.org/essays/output.cfm?file_id=3647 (accessed October 2007).

Wise, S. *Bringing Benefits Together: Capturing the Value(s) of Raindrops Where They Fall.* Presented at the U.S. EPA Wet Weather and CSO Technology Workshop, Florence, KY, September 2007a.

Wise, S. *Cities & Green Infrastructure: Examples from Chicago, Milwaukee, & Philadelphia.* Presented at U.S. EPA Wet Weather and CSO Technology Workshop, Florence, KY, September 2007b.

Yardley, W. Gaping Reminders of Aging and Crumbling Pipes. *New York Times*. February 8, 2007.

WASTEWATER UTILITY

Alabama
Montgomery Water Works & Sanitary Sewer Board

Alaska
Anchorage Water & Wastewater Utility

Arizona
Avondale, City of
Glendale, City of, Utilities Department
Mesa, City of
Peoria, City of
Phoenix Water Services Dept.
Pima County Wastewater Management
Safford, City of
Tempe, City of

Arkansas
Little Rock Wastewater Utility

California
Central Contra Costa Sanitary District
Corona, City of
Crestline Sanitation District
Delta Diablo Sanitation District
Dublin San Ramon Services District
East Bay Dischargers Authority
East Bay Municipal Utility District
El Dorado Irrigation District
Fairfield-Suisun Sewer District
Fresno Department of Public Utilities
Inland Empire Utilities Agency
Irvine Ranch Water District
Las Gallinas Valley Sanitary District
Las Virgenes Municipal Water District
Livermore, City of
Los Angeles, City of
Los Angeles County, Sanitation Districts of
Napa Sanitation District
Novato Sanitary District
Orange County Sanitation District
Palo Alto, City of
Riverside, City of
Sacramento Regional County Sanitation District
San Diego Metropolitan Wastewater Department, City of
San Francisco, City & County of
San Jose, City of
Santa Barbara, City of
Santa Cruz, City of
Santa Rosa, City of
South Bayside System Authority
South Coast Water District
South Orange County Wastewater Authority
Stege Sanitary District
Sunnyvale, City of
Union Sanitary District
West Valley Sanitation District

Colorado
Aurora, City of
Boulder, City of
Greeley, City of
Littleton/Englewood Water Pollution Control Plant
Metro Wastewater Reclamation District, Denver

Connecticut
Greater New Haven WPCA
Stamford, City of

District of Columbia
District of Columbia Water & Sewer Authority

Florida
Broward, County of
Fort Lauderdale, City of
Jacksonville Electric Authority (JEA)
Miami-Dade Water & Sewer Authority
Orange County Utilities Department
Pinellas, County of
Reedy Creek Improvement District
Seminole County Environmental Services
St. Petersburg, City of
Tallahassee, City of
Tampa, City of
Toho Water Authority
West Palm Beach, City of

Georgia
Atlanta Department of Watershed Management
Augusta, City of
Clayton County Water Authority
Cobb County Water System
Columbus Water Works
Fulton County
Gwinnett County Department of Public Utilities
Savannah, City of

Hawaii
Honolulu, City & County of

Idaho
Boise, City of

Illinois
Decatur, Sanitary District of
Greater Peoria Sanitary District
Kankakee River Metropolitan Agency
Metropolitan Water Reclamation District of Greater Chicago
Wheaton Sanitary District

Iowa
Ames, City of
Cedar Rapids Wastewater Facility
Des Moines, City of
Iowa City

Kansas
Johnson County Wastewater
Lenexa, City of
Unified Government of Wyandotte County/ Kansas City, City of

Kentucky
Louisville & Jefferson County Metropolitan Sewer District

Louisiana
Sewerage & Water Board of New Orleans

Maine
Bangor, City of
Portland Water District

Maryland
Anne Arundel County Bureau of Utility Operations
Howard County Bureau of Utilities
Washington Suburban Sanitary Commission

Massachusetts
Boston Water & Sewer Commission
Massachusetts Water Resources Authority (MWRA)
Upper Blackstone Water Pollution Abatement District

Michigan
Ann Arbor, City of
Detroit, City of
Holland Board of Public Works
Saginaw, City of
Wayne County Department of Environment
Wyoming, City of

Minnesota
Rochester, City of
Western Lake Superior Sanitary District

Missouri
Independence, City of
Kansas City Missouri Water Services Department
Little Blue Valley Sewer District
Metropolitan St. Louis Sewer District

Nebraska
Lincoln Public Works and Utilities Department

Nevada
Henderson, City of
Las Vegas, City of
Reno, City of

New Jersey
Bergen County Utilities Authority
Ocean County Utilities Authority
Passaic Valley Sewerage Commissioners

New York
New York City Department of Environmental Protection

North Carolina
Charlotte/Mecklenburg Utilities
Durham, City of
Metropolitan Sewerage District of Buncombe County
Orange Water & Sewer Authority
University of North Carolina, Chapel Hill

Ohio
Akron, City of
Butler County Department of Environmental Services
Columbus, City of
Metropolitan Sewer District of Greater Cincinnati
Montgomery, County of
Northeast Ohio Regional Sewer District
Summit, County of

Oklahoma
Oklahoma City Water & Wastewater Utility Department
Tulsa, City of

Oregon
Albany, City of
Clean Water Services
Eugene, City of
Gresham, City of
Portland, City of Bureau of Environmental Services
Water Environment Services

Pennsylvania
Hemlock Municipal Sewer Cooperative (HMSC)
Philadelphia, City of
University Area Joint Authority

South Carolina
Charleston Water System
Mount Pleasant Waterworks & Sewer Commission
Spartanburg Water

Tennessee
Cleveland Utilities
Murfreesboro Water & Sewer Department
Nashville Metro Water Services

Texas
Austin, City of
Dallas Water Utilities
Denton, City of
El Paso Water Utilities

WERF SUBSCRIBERS

Fort Worth, City of
Houston, City of
San Antonio Water System
Trinity River Authority

Utah
Salt Lake City Corporation

Virginia
Alexandria Sanitation Authority
Arlington, County of
Fairfax County
Hampton Roads Sanitation District
Hanover, County of
Henrico, County of
Hopewell Regional Wastewater Treatment Facility
Loudoun Water
Prince William County Service Authority
Richmond, City of
Rivanna Water & Sewer Authority

Washington
Everett, City of
King County Department of Natural Resources
Seattle Public Utilities
Sunnyside, Port of
Yakima, City of

Wisconsin
Green Bay Metro Sewerage District
Kenosha Water Utility
Madison Metropolitan Sewerage District
Milwaukee Metropolitan Sewerage District
Racine, City of
Sheboygan Regional Wastewater Treatment
Wausau Water Works

Australia
ACTEW (Ecowise)
South Australian Water Corporation
South East Water Limited
Sydney Water Corporation
Water Corporation of Western Australia

Canada
Edmonton, City of/Edmonton Waste Management Centre of Excellence
Lethbridge, City of
Regina, City of, Saskatchewan
Toronto, City of, Ontario
Winnipeg, City of, Manitoba

New Zealand
Watercare Services Limited

United Kingdom
Anglian Water Services, Ltd.

STORMWATER UTILITY

California
Fresno Metropolitan Flood Control District
Los Angeles, City of, Department of Public Works
Monterey, City of
San Francisco, City & County of
Santa Rosa, City of
Sunnyvale, City of

Colorado
Aurora, City of
Boulder, City of

Florida
Orlando, City of

Iowa
Cedar Rapids Wastewater Facility
Des Moines, City of

Kansas
Overland Park, City of

Kentucky
Louisville & Jefferson County Metropolitan Sewer District

Maine
Portland Water District

North Carolina
Charlotte, City of, Stormwater Services

Pennsylvania
Philadelphia, City of

Tennessee
Chattanooga Stormwater Management

Texas
Harris County Flood Control District, Texas

Washington
Bellevue Utilities Department
Seattle Public Utilities

STATE

Connecticut Department of Environmental Protection
Kansas Department of Health & Environment
New England Interstate Water Pollution Control Commission (NEIWPCC)
Ohio River Valley Sanitation Commission
Urban Drainage & Flood Control District, CO

CORPORATE

ADS Environmental Services
Advanced Data Mining International
AECOM
Alan Plummer & Associates
Alpine Technology Inc.
Aqua-Aerobic Systems Inc.
Aquateam–Norwegian Water Technology Centre A/S
ARCADIS
Associated Engineering
Bernardin Lochmueller & Associates
Black & Veatch
Blue Water Technologies, Inc.
Brown & Caldwell
Burgess & Niple, Ltd.
Burns & McDonnell
CABE Associates Inc.
The Cadmus Group
Camp Dresser & McKee Inc.
Carollo Engineers Inc.
Carpenter Environmental Associates Inc.
CET Engineering Services
CH2M HILL
CRA Infrastructure & Engineering
CONTECH Stormwater Solutions
D&B/Guarino Engineers, LLC
Damon S. Williams Associates, LLC
Ecovation
EMA Inc.
Environmental Operating Solutions, Inc.
Environ/The ADVENT Group, Inc.
Fay, Spofford, & Thorndike Inc.
Freese & Nichols, Inc.
ftn Associates Inc.
Gannett Fleming Inc.
Garden & Associates, Ltd.
Geosyntec Consultants
GHD
Greeley and Hansen LLC
Hazen & Sawyer, P.C.
HDR Engineering Inc.
HNTB Corporation
Hydromantis Inc.
HydroQual Inc.
Infilco Degremont Inc.
Jacques Whitford NAWE, Inc.
Jason Consultants LLC Inc.
Jordan, Jones, & Goulding Inc.
KCI Technologies Inc.
Kelly & Weaver, P.C.
Kennedy/Jenks Consultants
KMK Consultants
Larry Walker Associates
Limno-Tech Inc.
The Low Impact Development Center Inc.
Malcolm Pirnie Inc.
Material Matters
McKim & Creed
MPR Engineering Corporation, Inc.
MWH
NTL Alaska, Inc.
O'Brien & Gere Engineers Inc.
Odor & Corrosion Technology Consultants Inc.
Paradigm Environmental Technologies, Inc.
Parametrix Inc.
Parsons
Post, Buckley, Schuh & Jernigan
Praxair, Inc.
Ring Industrial Group
RMC Water & Environment
Ross & Associates Ltd.
SAIC
Siemens Water Technologies
The Soap & Detergent Association
Smith & Loveless, Inc.
Southeast Environmental Engineering, LLC
Stearns & Wheler, LLC
Stone Environmental Inc.
Stratus Consulting Inc.
Synagro Technologies Inc.
Tetra Tech Inc.
Trojan Technologies Inc.
Trussell Technologies, Inc.
Uni-Bell PVC Pipe Association
URS Corporation
Westin Engineering Inc.
Wright Water Engineers
Zoeller Pump Company

INDUSTRY

American Electric Power
American Water
Anglian Water Services, Ltd.
Chevron Energy Technology
The Coca-Cola Company
Dow Chemical Company
DuPont Company
Eastman Chemical Company
Eli Lilly & Company
Johnson & Johnson
Merck & Company Inc.
Procter & Gamble Company
Suez Environnment
United Utilities North West (UUNW)
United Water Services LLC

Note: List as of 2/21/09

IWA Publishing's authorised EU representative for General Product Safety Regulations is Diane D'Arras, 15 rue Duret, 75116 Paris, France, e-mail: safety@iwap.co.uk.

Printed and bound by CPI Group (UK) Ltd, Croydon, CR0 4YY
12/05/2026
02108892-0013